FORD SIERRA

1982 to 1992

All Ford Sierra models (including Sapphire) except RS Cosworth derivatives, P100 Pick-up and Diesel engine models.

HANDBOOK & DRIVERS GUIDE

by Steve Rendle

Full details of all servicing and repair tasks for the models covered by this Handbook can be found in the relevant Owners Workshop Manual – OWM **903** Ford Sierra 4-cylinder, OWM **904** Ford Sierra V6.

(1813 – 1W1)

A Haynes Handbook and Drivers Guide

Printed and published by J H Haynes & Co Ltd Sparkford Nr Yeovil Somerset BA22 7JJ England

ISBN 1 85010 813 7

British Library Cataloguing in Publication Data. A catalogue record of this book is available from the British Library

ACKNOWLEDGEMENTS

We gratefully acknowledge the assistance of RoSPA in compiling the information used in this Handbook. Thanks are also due to Duckhams Oils who provided lubrication data and to Britax who supplied information on child safety. Certain illustrations, including the cover photography, are the copyright of the Ford Motor Company Limited and are used with their permission. Additional photographs were supplied by Quadrant Picture Library/Auto Express. Thanks are also due to all those people at Sparkford who helped in the production of this Handbook.

We take great pride in the accuracy of information given in this Handbook, but vehicle manufacturers make alterations and design changes during the production run of a particular vehicle of which they do not inform us. No liability can be accepted by the authors or publishers for loss, damage or injury caused by any errors in, or omissions from, the information given.

CONTENTS

About this Handbook	4
The Sierra family	7
Buying and selling	9
General tips on buying	9
Points to look for when buying	10
General tips on selling	11
Dimensions and weights	13
Controls and equipment	15
Driver's instruments and controls	19
Interior equipment	29
Exterior equipment	32
Accidents and emergencies	34
How to cope with an accident	35
First aid	36
Requirements of the law	38
Essential details to record	38
Check list and accident report form	39
How to cope with a fire	42
How to cope with a broken windscreen	42
What to do if your car is broken into	42
Breakdowns	43
Breakdowns on an ordinary road	43
Motorway breakdowns	44
Changing a wheel	44
Towing	46
Starting a car with a flat battery	47
What to carry in case of a breakdown	48
Driving safety	49
Before starting a journey	49
Driving in bad weather	50
Motorway driving	51
Towing a trailer or caravan	52
Alcohol and driving	54
Skid control	54
Advice to women drivers	55
Child safety	57
Driving Abroad	59
Reducing the cost of motoring	63
Car crime prevention	65
Service specifications	67
Regular checks	71
Checking oil level	74
Checking coolant level	75
Checking brake fluid level	76
Checking tyres	76
Checking washer fluid level	78
Checking battery electrolyte level	78
Checking wipers and washers	78
Checking lights and horn	79
Checking for fluid leaks	79
Servicing	81
Service schedule	82
Safety first!	86
Buying spare parts	89
Service tasks	95
Seasonal servicing	114
Tools	116
What to buy	116
Care of your tools	116
Bodywork and interior care	117
Cleaning the interior	117
Cleaning the bodywork	117
Dealing with scratches	118
Bulb, fuse and relay renewal	119
Bulbs – exterior lights	120
Bulbs – interior lights	125
Fuses	126
Relays	128
Preparing for the MOT test	129
Fault finding	131
Car jargon	137
Local radio frequencies	147
Conversion factors	152
Distance Tables	154
Index	156

ABOUT THIS HANDBOOK

▲ *Sierra XR4x4 2.9 (1989)*

The idea behind this Handbook is to help you to get the most out of your motoring.

Apart from the things that every owner needs to know, to deal with unexpected mishaps like a puncture or a blown light bulb, you'll find clearly presented information on road safety, hints on driving abroad, and tips on how to prepare your car for the MOT test. We've also included details of local radio frequencies to help you avoid those inevitable 'jams', and there's a Section on what to do in the unfortunate event of an accident.

For those not familiar with their car, there's a Section to explain the location and the operation of the various controls and instruments. Additionally, there's a Section on fault finding, and a useful glossary of car jargon.

Garage labour charges usually form the major part of any car servicing bill, and we hope to help you to reduce those bills by carrying out the more straightforward routine servicing jobs yourself. If you're about to start carrying out your own servicing for the first time, we aim to provide you with easy-to-follow instructions, enabling you to carry out the simpler tasks which perhaps you've left to a garage or a 'car-minded' friend in the past. Even if you prefer to leave regular servicing to a suitably qualified expert, by using this book you'll be able to carry out regular checks on your car to make sure that your motoring is safe and hopefully trouble-free. You'll also find advice on buying suitable tools, and safety in the home workshop.

Some readers of this Handbook may not yet have bought a Sierra, so we've included a brief history of the range, and some useful tips on buying and selling.

All in all, we hope that this book will prove a handy companion for your motoring adventures, and hopefully we'll help to reduce the problems which inevitably crop up in everyday driving. If you're bitten by the DIY bug, and you're keen to tackle some of the more advanced repair jobs on your car, then you'll need our **Owners Workshop Manual** for your particular model (OWM 903 for Sierra 4-cylinder engine models, or OWM 904 for Sierra V6 engine models). These manuals give a step-by-step handbook to all the repair and overhaul tasks, with plenty of illustrations to make things even clearer.

Happy motoring!

THE SIERRA FAMILY

▲ *Sierra 1.6 L Hatchback (1983)*

▲ *Sierra 2.0 GL Estate (1985)*

▲ *Sierra Sapphire 1.8 LX (1989)*

The Sierra was first introduced into the UK in late 1982 as the successor to the famous Cortina, and was one of the first cars to make use of the 'aeroback' body style, designed to reduce the air drag and improve fuel economy.

To begin with, only Hatchback and Estate bodystyles were available, but the range was gradually developed to include a vast variety of different models and equipment. In 1987, the whole range was facelifted and heavily revised to incorporate many improvements over the original design. At the same time the Saloon models, known as Sapphires, were launched. Models have been produced with 1.3, 1.6, 1.8, 2.0, 2.3, 2.8 & 2.9 litre engines, 4- & 5-speed manual gearboxes or 3- & 4-speed automatic transmissions, and two-wheel or four-wheel-drive. A large number of different trim levels have been available over the years, from the Base Hatchback, Saloon and Estate, to the performance-orientated XR4i and XR4x4 Hatchback models.

All members of the Sierra family share a similar mechanical layout, and they all use the same large family of engines and gearboxes, although many detailed modifications and improvements have been made since the original design was introduced.

Most of the mechanical components of the cars are fairly conventional, and the main systems are designed to keep servicing time and costs as low as possible. The electronic ignition system, clutch, brakes, wheel bearings, etc are all routine-maintenance-free, and should only require attention when certain items need to be renewed.

On all models, the engine is mounted at the front of the car, with the gearbox positioned behind the engine. On four-wheel-drive models, an additional 'transfer' gearbox is mounted behind the main gearbox to provide drive to all four wheels.

A wide range of standard and optional equipment is available across the model ranges, including items such as sunroofs, electric windows, central door locking and, on certain models, anti-lock brakes.

With the wide availability of spare parts, these cars are ideal for the enthusiastic amateur mechanic, who wants to keep running costs to a minimum.

▲ *Sierra XR4i (1985)*

▲ *Sierra XR4i and XR4x4*

▲ *Sierra LX*

GENERAL TIPS ON BUYING

- **Don't rush out and buy the first car to catch your attention**
- **Always buy from a recognised dealer in preference to buying privately**
- **Have the car checked over by someone knowledgeable before you buy it**
- **In general, you get what you pay for!**

Before buying a second-hand car, it's worthwhile doing some homework to try and avoid some of the pitfalls waiting for the unwary. First of all, don't rush out and buy the first car to catch your attention (all that glitters is not gold!), and remember that much of the responsibility is yours when it comes to the soundness of the deal, especially when buying a car privately.

Wherever possible, buy from a recognised dealer, and check that the dealer is a member of the Retail Motor Industry Federation, as this will provide you with certain legal safeguards if you have any problems. If you buy from a dealer, you are covered by the Sale Of Goods Act, which in summary states that the goods must be fit for their intended purpose, the goods must be of proper quality, and the goods must be as described by the seller. If you're buying a car privately, ask to see the service receipts and the MOT certificates going back as long as the car has been in the possession of the current owner (this will help to establish that the car hasn't been stolen, and that the recorded mileage is genuine), and always ask to view the car at the seller's private address (to make sure that the car isn't being sold by an unscrupulous dealer posing as a private seller). Check the vehicle documents for obvious signs of forgery and, if in doubt, contact the DVLA and give them details of the Registration Document, as they will be able to run a check on its authenticity.

As far as the soundness of the car itself is concerned, a genuine service history is helpful. This is provided by the service book supplied with the car when new, which should be completed and officially stamped by an authorised garage after each service. Cars with a full service history (fsh) usually command a higher price than those without.

To check the condition of a car, a professional examination is well worthwhile, if you can afford it, and organisations such as the AA and RAC will be able to provide such a service. Otherwise, you must trust your own judgement, and/or that of a knowledgeable friend. Although it's tempting, try not to overlook the mechanical soundness of the car in favour of the overall appearance. It's relatively easy to clean and polish a car every week, but when were the brakes and tyres last checked? Above all, safety must always take priority. Don't view a car in the wet, as water on the bodywork can give a misleading impression of the condition of the paintwork. First of all, check around the outside of the car for rust, and for obvious signs of new or mis-matched paintwork which might show that the car has been involved in an accident. Check the tyres for signs of unusual wear or damage, and check that the car 'sits' evenly on its suspension, with all four corners at a similar height. Open the bonnet and check for any obvious signs of fluid leakage (oil, water, brake fluid), then start the engine and listen for any unusual noises – some background noise is to be expected on older cars, but there should be no sinister rattles or bangs! Also listen to the exhaust to make sure that it isn't 'blowing' indicating the need for renewal, and check for signs of excessive exhaust smoke. Black smoke may be caused by poorly adjusted fuel mixture, which can usually be rectified fairly easily, but blue smoke usually indicates worn engine components, which may prove expensive to repair. Finally, drive the car, and test the brakes, steering and gearbox. Make sure that the car doesn't pull to one side, and check that the steering feels positive and that the gears can be selected satisfactorily without undue harshness or noise. Listen for any unusual noises or vibration, and keep an eye on the instruments and warning lights to make sure that they are working and indicating correctly.

If all proves satisfactory, try to negotiate a suitable deal, but remember that a dealer has to work to a profit margin, and it's unlikely that you'll find a good car for a silly price. Always obtain a receipt for your money.

On the whole, it's true to say that you get what you pay for. Above all, don't be rushed into making a hasty decision.

POINTS TO LOOK FOR WHEN BUYING A SIERRA

In addition to the points outlined previously, there are a few special points to look for when buying a Sierra.

In general, the later models built from 1987 are an all-round improvement on the earlier models, and will prove a better buy if your budget allows.

Corrosion is not usually a problem on Sierras, but there are a few areas which are prone to trouble.

Firstly, check the wheel arches behind each front wheel. Flying stones can damage the underseal, and this leads to rust. Turn the steering wheel to full lock to allow access, and probe the wheel arches with your finger or a screwdriver.

Other areas prone to rust are the bottom rear edges of the doors, and the lower edge of the tailgate on Hatchback models. Also check the windscreen pillars, and on Estate models the curved panel under the rear bumper.

Check the bumpers for any signs of damage or cracks, as these components are expensive to renew.

Check carefully for accident damage at the front and rear of the car, especially on early models. Look behind the front bumpers, and check for signs of corrosion, damage or repairs to the body panels. On Hatchback models, check for ridges or other distortion in the floor of the luggage compartment (lift up the carpet to check).

The Sierra is a popular company car, and is also a target for 'clocking' by unscrupulous dealers. Check the interior for signs of unusual wear (worn driver's carpet, steering wheel, pedal rubber pads etc) which may indicate that the recorded mileage is not genuine. Also check the facia around the instruments for any marks or damage which may indicate that the speedometer has been tampered with.

After high mileages, all the engines used in the Sierra are prone to a build-up of black sludge inside the engine if the oil is not changed regularly. To check for this, remove the oil filler cap from the top of the engine, and run a finger around the inside of the hole. If sludge is present, reject the car.

High-mileage cars, especially early 2.0 litre models, are prone to engine camshaft wear, which can be expensive to cure. A worn camshaft will cause a loud rattle when the engine is running, and the rattle will become louder as the engine is 'revved'.

Another point to note on OHC and CVH engines is that the camshaft drivebelt (timing belt) should be renewed every 36 000 miles. If you're not sure whether this has been done, it's sensible to have the necessary work carried out as soon as you buy the car, as if the drivebelt breaks in service, very serious and expensive engine damage can occur.

Unfortunately Sierras are a common target for thieves, so check the interior trim, particularly around the radio, for signs of damage.

If possible, try to avoid 1.3 & 2.3 litre models. 1.3 litre models suffer from poor performance (they are rather underpowered), while 2.3 litre models give poor economy and very little extra performance over 2.0 litre models.

Sierras are plentiful on the secondhand market (although low-mileage examples are rare), and it shouldn't be difficult to find a good car which will give trouble-free service.

GENERAL TIPS ON SELLING

- **Make sure that the car is clean and tidy**
- **Make sure that all of the service documents, registration document etc, are available for inspection**
- **Ask yourself ... 'Would I buy this car?'**

Obviously when selling a car, bear in mind the points which the prospective buyer should be looking for, as described in the above sections. It goes without saying that the car should be clean and tidy, as first impressions are important. Any fluid leaks should be cured, and there's no point in trying to disguise any major bodywork or mechanical problems.

If you're trading the car in with a dealer, you will always get a lower price than if you sell privately, but you can be fairly sure that there will be less comeback to you should any unexpected problems develop. If selling privately, don't allow the buyer to take the car away until you have his/her money, and it's a good idea to ask him/her to sign a piece of paper to say that he/she is happy to buy the car as viewed, just in case any problems develop later on. Give a receipt for the money paid.

DIMENSIONS AND WEIGHTS

Note: *All figures are approximate, and will vary depending on model*

DIMENSIONS [mm (in)]

Overall length

Saloon (Sapphire) models	**4467** (176.0)
Hatchback models	**4407** to **4460** (173.6 to 175.7)
Estate models	**4506** to **4522** (177.5 to 178.2)

Overall width

All models	**1820** to **1920** (71.7 to 75.6)

Overall height

Saloon (Sapphire) models	**1352** to **1407** (53.3 to 55.4)
Hatchback models	**1352** to **1420** (53.3 to 55.9)
Estate models	**1386** to **1506** (54.6 to 59.3)

WEIGHTS [kg (lb)]

Nominal kerb weight

Saloon (Sapphire) models	**1025** to **1135** (2260 to 2503)
Hatchback models:	
All except V6 engine models	**1010** to **1145** (2227 to 2525)
V6 engine models	**1165** to **1290** (2569 to 2844)
Estate models:	
All except V6 engine models	**1065** to **1185** (2348 to 2613)
V6 engine models	**1200** to **1315** (2646 to 2900)

Maximum roof rack load

All models	**75** (165)

Maximum towing weight (for trailer or caravan)

All except 1.3 and 1.6 litre models	**1350** (2977)
1.3 litre models	**825** (1819)
1.6 litre models (except Estate)	**1100** (2426)
1.6 litre Estate models	**1250** (2756)

Maximum trailer/caravan noseweight

All models	**50** (110)

CONTROLS AND EQUIPMENT

For those not familiar with the Sierra models, this Section will help to identify the instruments and controls. Typical instrument panel layouts are shown in the accompanying illustrations. The operation of most equipment is self-explanatory, but some items require further explanation to ensure that their use is fully understood.

Note that not all items are fitted to all models.

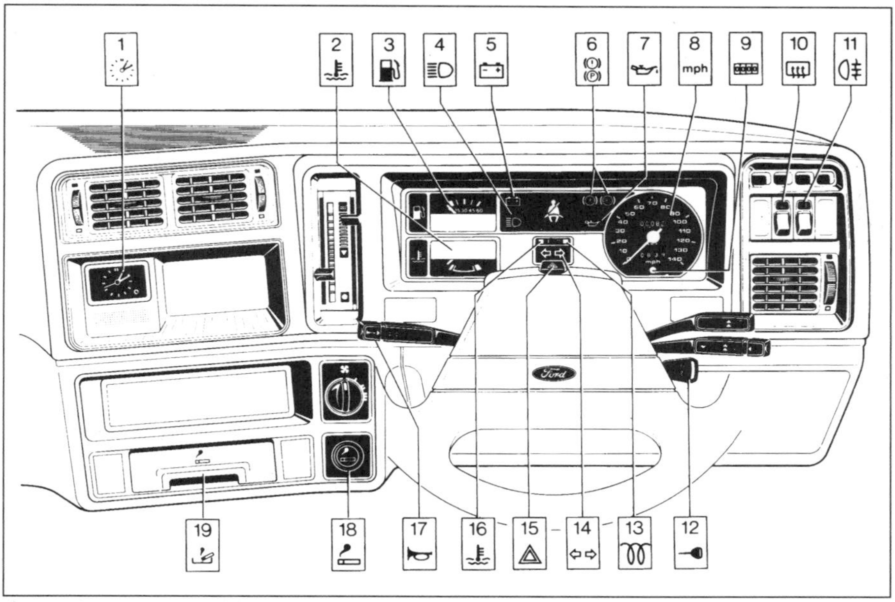

▲ *Typical instrument panel for 'low series' Sierra models up to 1987*

1 *Clock*
2 *Temperature gauge*
3 *Fuel gauge*
4 *Headlight main beam warning light*
5 *Ignition warning light*
6 *Handbrake 'on'/low brake fluid level warning light*
7 *Low oil pressure warning light*
8 *Speedometer*
9 *Mileage recorder reset button*
10 *Heated rear window switch*
11 *Rear foglight switch*
12 *Ignition switch/steering lock*
13 *Glow plug warning light (Diesel models only)*
14 *Direction indicator warning light*
15 *Hazard flasher switch*
16 *Temperature warning light (Diesel models only)*
17 *Horn switch*
18 *Cigarette lighter*
19 *Ashtray*

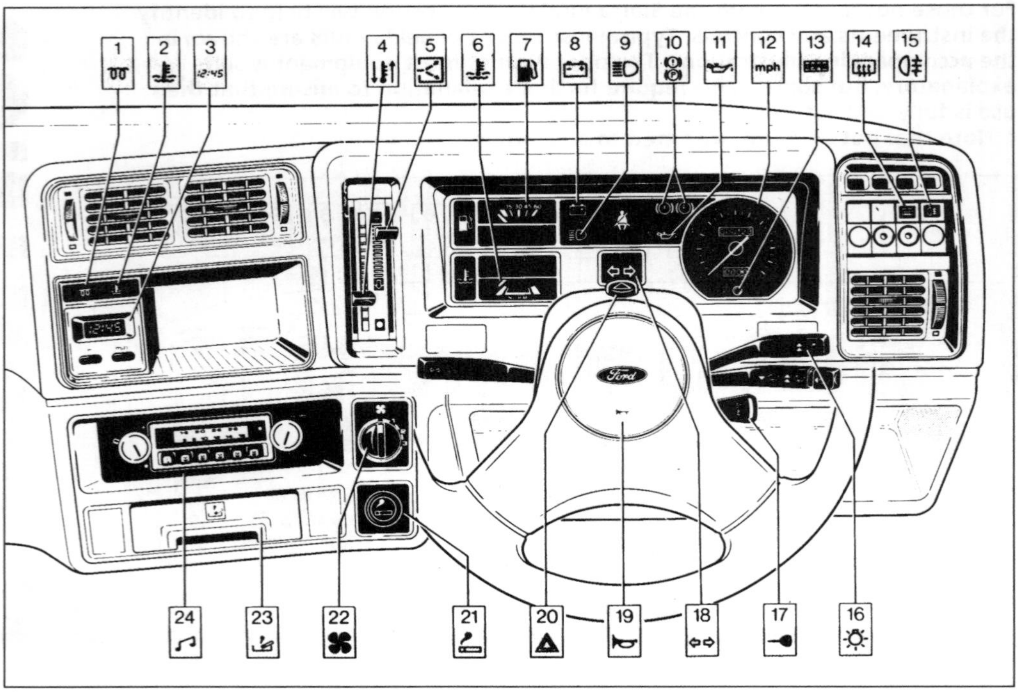

▲ *Typical instrument panel for 'low series' Sierra models from 1987*

1 *Glow plug warning light (Diesel models only)*
2 *Temperature warning light (Diesel models only)*
3 *Clock*
4 *Air temperature control lever*
5 *Air distribution control lever*
6 *Temperature gauge*
7 *Fuel gauge*
8 *Ignition warning light*
9 *Headlight main beam warning light*
10 *Handbrake 'on'/low brake fluid level/ABS warning light*
11 *Low oil pressure warning light*
12 *Speedometer*
13 *Mileage recorder reset button*
14 *Heated rear window switch*
15 *Rear foglight switch*
16 *Exterior lights switch*
17 *Ignition switch/steering lock*
18 *Direction indicator warning light*
19 *Horn switch*
20 *Hazard flasher switch*
21 *Cigarette lighter*
22 *Heater booster fan switch*
23 *Ashtray*
24 *Radio*

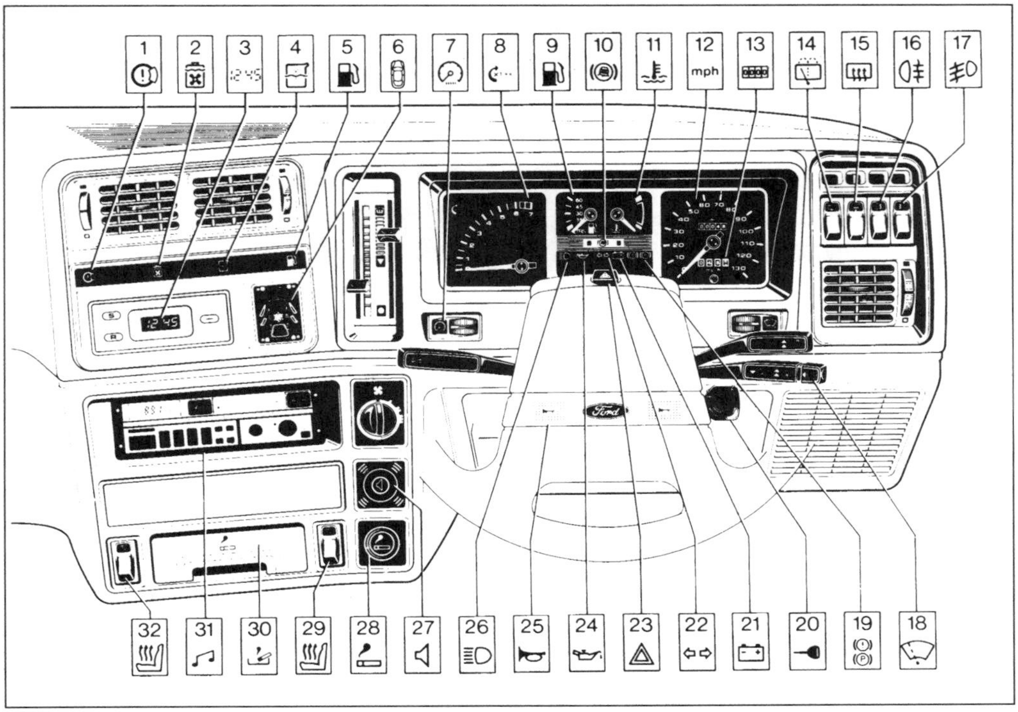

▲ *Typical instrument panel for 'high series' Sierra models up to 1987*

1 *Brake pad wear warning light*
2 *Low coolant level warning light*
3 *Clock*
4 *Low washer fluid level warning light*
5 *Low fuel level warning light*
6 *Graphic warning display*
7 *Instrument illumination control*
8 *Tachometer ('Rev. counter')*
9 *Fuel gauge*
10 *ABS warning light*
11 *Temperature gauge*
12 *Speedometer*
13 *Mileage recorder reset button*
14 *Tailgate wash/wipe switch*
15 *Heated rear window switch*
16 *Rear foglight switch*
17 *Front foglight switch*
18 *Windscreen wash/wipe switch*
19 *Handbrake 'on'/low brake fluid level/ABS warning light*
20 *Ignition switch/steering lock*
21 *Ignition warning light*
22 *Direction indicator warning light*
23 *Hazard flasher switch*
24 *Low oil pressure warning light*
25 *Horn switch*
26 *Headlight main beam warning light*
27 *Loudspeaker balance control*
28 *Cigarette lighter*
29 *Heated driver's seat switch*
30 *Ashtray*
31 *Radio*
32 *Heated front passenger's seat switch*

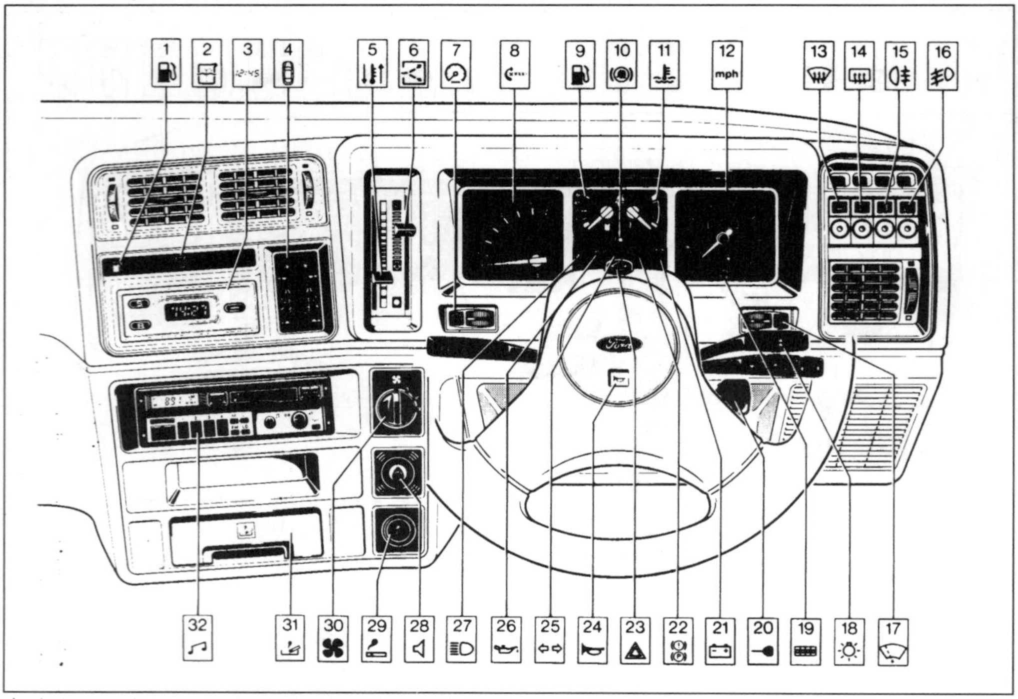

▲ *Typical instrument panel for 'high series' Sierra models from 1987*

1 *Low fuel level warning light*
2 *Low washer fluid level warning light*
3 *Clock*
4 *Graphic warning display*
5 *Air temperature control lever*
6 *Air distribution control lever*
7 *Instrument illumination control*
8 *Tachometer ('Rev. counter')*
9 *Fuel gauge*
10 *ABS warning light*
11 *Temperature gauge*
12 *Speedometer*
13 *Heated windscreen switch*
14 *Heated rear window switch*
15 *Rear foglight switch*
16 *Front foglight switch*
17 *Variable intermittent wipe control*
18 *Exterior lights switch*
19 *Mileage recorder reset button*
20 *Ignition switch/steering lock*
21 *Ignition warning light*
22 *Handbrake 'on'/low brake fluid level/ABS warning light*
23 *Hazard flasher switch*
24 *Horn switch*
25 *Direction indicator warning light*
26 *Low oil pressure warning light*
27 *Headlight main beam warning light*
28 *Loudspeaker balance control*
29 *Cigarette lighter*
30 *Heater booster fan switch*
31 *Ashtray*
32 *Radio*

DRIVER'S INSTRUMENTS AND CONTROLS

Speedometer

Indicates the car's road speed in miles per hour (mph). Also incorporates a mileage recorder, and a separate trip meter which can be reset by pressing the button protruding from the instrument.

Tachometer ('Rev. counter')

Indicates engine speed in revolutions per minute (rpm). For normal driving and the best fuel economy, the engine speed should be kept in the 2000 to 4000 rpm range. The red band indicates the maximum permissible engine speed.

Fuel gauge

Indicates the quantity of fuel remaining in the tank. When the needle enters the red band, there are approximately 8 litres (1.8 gallons) of fuel remaining.

Temperature gauge

Indicates the engine coolant temperature. As the engine warms up, the needle should move from the blue (cold) end of the scale into the middle region. If the needle remains in the blue section, or enters the red (hot) section, a fault is indicated and advice should be sought (refer to *'Fault finding'* on page 131).

Do not continue to run an engine which shows signs of overheating

Clock

Four different types of clock have been used.

ANALOGUE TYPE

To adjust the time, push in the button and turn it.

DIGITAL TYPE

To adjust the time, switch on the ignition. Press the button on the left to change the hour setting, and press the button on the right to change the minutes setting (note that some early models may have recessed buttons which must be pressed using a pointed instrument such as a ballpoint pen). To synchronise the clock with an exact time signal, press and hold both buttons. Release the buttons at the time signal, and set the time as described previously. On early models, the clock displays the time in the 12-hour format, but on later models the time can be displayed in the 12- or 24-hour format. To set the time format on later models, press both buttons simultaneously and then release them – if the display shows 1:00, the clock is in the 12-hour mode, and if the display shows 0:00 the clock is in the 24-hour mode.

MULTI-FUNCTION DIGITAL TYPE

The clock can display the date, and has a timer feature. No display is shown with the ignition switched off, but the display can be recalled for 4 seconds by pressing the 'select' button **(A)**. One press will display time, and further presses will display the date and timer. The ignition must be switched on to carry out all setting and adjustments.

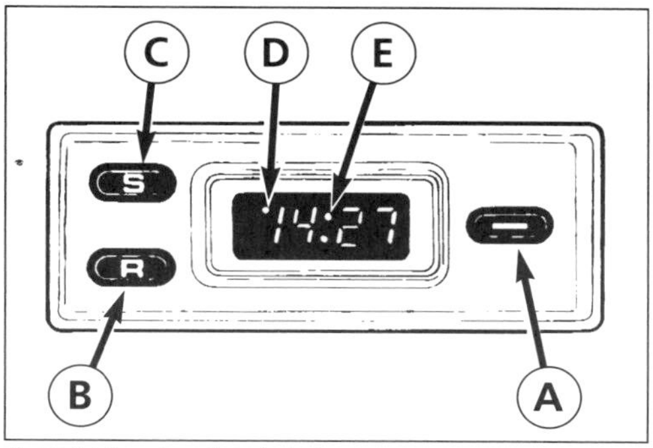

▲ *Multi-function digital clock*
- ***A*** *Select button*
- ***B*** *Reset button*
- ***C*** *Stop/start button*
- ***D*** *Timer symbol*
- ***E*** *Colon*

Adjusting time and date

The time and date must be set using the following sequence. With the display showing the time, press the 'reset' button **(B)** once. The display should flash, showing '12H' or '24H' depending on whether the display has been set to the 12- or 24-hour format. Press the 'stop/start' button **(C)** to change the format. Press the 'reset' button again. The hour number on the left will flash. If the 24-hour format was selected, the minute number on the right will remain stationary, but if the 12-hour format was selected, an 'A' for am or 'P' for pm will appear in place of the minutes. Press the 'stop/start' button to increase the hour number as required. Press the 'reset' button again. The minute number on the right of the display will now flash. Press the 'stop/start' button to increase the minute number. Press the 'reset' button again. The date will now be displayed, with the day number flashing on the left, and the month number stationary on the right. Press the 'stop/start' button to increase the day number. Press the 'reset' button again. The display now shows the date with the day stationary on the left, and the month number flashing on the right. Press the 'stop/start' button to increase the month number. When the adjustments are complete, press the 'reset' button to return to the time display. To return to the time display at any point during the procedure, press the 'select' button.

Date display

To display the date, press the 'select' button **(A)** once. The date will be displayed for 4 seconds, after which the time will reappear.

Timer

To operate the timer, with the display showing the time, press the 'select' button **(A)** twice. The timer can now be started by pressing the 'stop/start' button **(C)**. Pressing the 'stop/start' button again will stop the timer and freeze the display. A third press will restart the timer from where it stopped. The timer will count to 59 minutes 59 seconds, and then roll over to 1 hour 00 minutes and continue to count at one minute intervals. When the timer reaches 59 hours 59 minutes, it automatically resets to 0 and continues counting.

Either of the two other displays can be selected when the timer is working. To display the time, press the 'select' button once, and to display the date, press the 'select' button again. To reset the timer to zero with the clock showing time, press the 'select' button twice to display the timer, then to zero the timer press the 'reset' button **(B)**. The display can now be set to show the normal time by pressing the 'select' button, or the timer can be restarted by pressing the 'stop/start' button. The recorded time when the timer was stopped will be stored in the memory until the timer is restarted again, even if the ignition is switched off. If the ignition is switched off when the timer is still running, it will continue to count until the ignition is switched on again and the timer is stopped or reset; the timer can therefore be used to record actual journey time.

COMBINED ANALOGUE/MULTI-FUNCTION DIGITAL TYPE

The clock can display the date, and has alarm, stop-watch and timer functions. The functions are controlled by three buttons, and indicator lights below the digital display indicate which function has been selected.

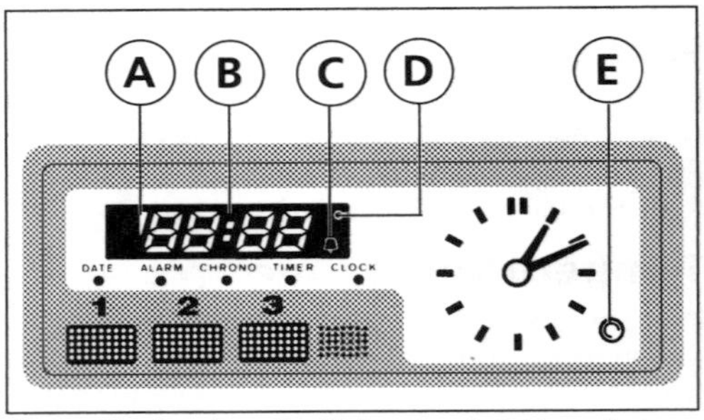

▲ *Combined analogue/multi-function digital type clock*
A *Seconds indicator*
B *Colon*
C *Alarm symbol*
D *Timer symbol*
E *Analogue clock adjustment knob*

Adjusting time ('CLOCK')

The digital clock displays two figures; hours/minutes followed by minutes/seconds. Each figure is shown for 4 seconds; then, provided no adjustments are made, the date is displayed. The analogue clock will not operate unless the digital clock has been started, as both clocks are directly coupled. To set the time on the digital clock, proceed as follows.

Press button **(1)** to select 'CLOCK' (clock indicator light illuminates). The display should show hours/minutes. Press button **(2)** for fast set, or button **(3)** for slow set (note that 4 seconds after any button was last pressed, the display will return to 'DATE').

Press button **(1)** again to display minutes/seconds, then press button **(2)** or **(3)** which will stop both the digital and analogue clocks, and zero the seconds.

Set the time on the analogue clock using the knob **(E)**.

To start both clocks, press button **(2)** or **(3)**. The display will show minutes/seconds for 4 seconds, then the date will be displayed.

Adjusting date ('DATE')

With the ignition switched on, press button **(2)** for fast set, or **(3)** for slow set until the desired date is displayed.

Alarm ('ALARM')

Press button **(1)** to select 'ALARM' (alarm indicator light illuminates). The time when the alarm will operate will be displayed in the 24-hour format.

To adjust the time, press button **(2)** for fast set, or **(3)** for slow set until the desired time is displayed. The alarm symbol **(C)** will appear in the bottom right-hand corner of the display. If no further adjustments are made within 4 seconds, the display will revert to 'DATE'.

To activate the alarm, look to see if the alarm symbol **(C)** is displayed. If it isn't, press button **(1)** again, and allow the display to revert to 'DATE', when the alarm symbol **(C)** should be displayed.

To cancel the alarm, press button **(1)**, and allow the display to revert to 'DATE' when the alarm symbol **(C)** should disappear.

To cancel the alarm buzzer when it sounds, press button **(1)**. If the button is pressed twice, the alarm will be cancelled, and the alarm symbol **(C)** will disappear. If the button is pressed once, the buzzer will stop, but will sound again every 10 minutes until the button is pressed a second time to cancel the alarm.

Stopwatch ('CHRONO')

Press button **(1)** to select 'CHRONO' (chrono indicator light illuminates). The display will show this mode until another mode is selected.

Press button **(3)** to stop or reset the stopwatch.

Press button **(2)** to start the stopwatch. The flashing colon **(B)** will indicate that the stopwatch is running.

Press button **(2)** again to 'freeze' the display. The flashing colon **(B)** will indicate that the stopwatch is still running.

Press button **(2)** again to 'unfreeze' the display and show the current time count.

Press button **(3)** to stop the stopwatch, then button **(1)** to select another function.

Timer ('TIMER')

Press button **(1)** to select 'TIMER' (timer indicator light illuminates). The display will show this mode until another mode is selected.

Press button **(3)** to stop or reset the timer.

Press button **(3)** again to start the timer. The display will show the total driving time.

Press button **(2)** to display the total journey time.

Press button **(2)** again to revert to total driving time.

The timer will continue to count when the ignition is switched off, to indicate total journey time, until the button **(3)** is pressed to reset the timer. The timer indicator light will be illuminated whenever the timer is running.

Fuel computer

Provides information on fuel level, distance and economy. There is also an audible warning to indicate low fuel level. The computer is controlled by three pushbuttons, and a tone will sound to indicate that a button has been pressed. The 'function' button **(D)** is used to switch between the four functions of the computer, and indicator lights show which function has been selected.

INSTANT ECONOMY

Shows the instantaneous fuel consumption, and this figure will change rapidly as driving conditions change during acceleration, braking, etc.

AVERAGE ECONOMY

Shows the average fuel consumption since the function was last reset.

FUEL USED

Shows the quantity of fuel used since the function was last reset.

RANGE

Gives an estimate of the distance the car will

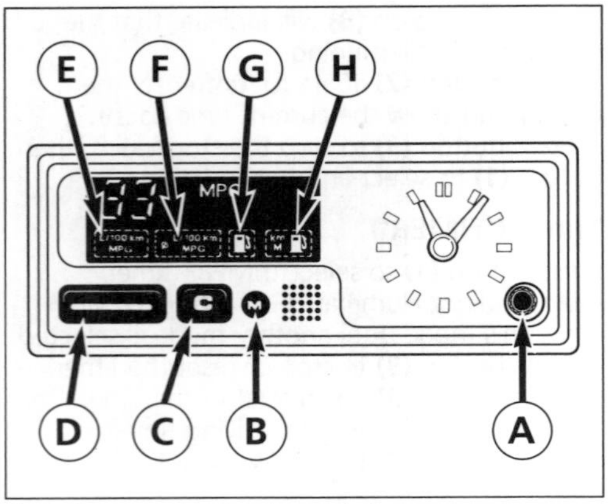

▲ *Fuel computer and analogue clock*
A *Analogue clock reset knob*
B *Mode button*
C *Clear button*
D *Function button*
E *'Instant economy' indicator light*
F *'Average economy' indicator light*
G *'Fuel used' indicator light*
H *'Range' indicator light*

travel on the fuel remaining in the tank. To remind the driver that the fuel level is low, an audible warning will sound when the range decreases to 50, 25 and 10 miles (80, 40 and 20 km) respectively. To stop the warning, any of the three buttons can be pressed, but in any case, the warning will stop automatically after sounding five times. Approximately 7 litres (1.5 gallons) of fuel are left in the tank when the range indicates 0. The computer will automatically calculate a new range when more than 9 litres (2 gallons) of fuel are added to the tank.

When the ignition is switched on, the computer shows the last selected display. However, if the range is less than 50 miles (80 km), the audible warning will sound and the range will be displayed to remind the driver that the fuel level is low. Pressing the 'clear' button **(C)** while in the 'Fuel used' function sets the value to 0. Pressing the 'clear' button while in the 'Average economy' function sets the value to the current 'Instant economy' value. Pressing the 'mode' button **(B)** changes the units displayed from metric to imperial, and *vice-versa*.

Trip computer

Certain early models were fitted with a trip computer which can provide extensive information on fuel, speed, distance and time, and which has a number of alarm and warning functions.

The trip computer is a complex piece of equipment, and a full description of its operation is beyond the scope of this Handbook. Full details of the trip computer can be found in a special booklet produced by Ford, which should have been supplied with the car when new.

Warning lights

These lights warn the driver of a fault, or inform the driver that a particular device is in operation.

HANDBRAKE 'ON'/LOW BRAKE FLUID LEVEL/ABS WARNING LIGHT

Warns that the handbrake is applied, or the brake fluid level is low. If the light comes on while driving, and the handbrake is fully released, stop immediately and check the brake fluid level (refer to *'Regular checks'* on page 70). On models fitted with an Anti-lock Braking System (ABS), this light has an extra function, and may illuminate in conjunction with the ABS warning light. If this happens, be prepared for the possibility of increased brake pedal travel and increased stopping distance, and refer the problem to a Ford dealer as soon as possible.

Do not continue to drive the car if a brake fluid leak is suspected

LOW OIL PRESSURE WARNING LIGHT

Warns that the engine oil pressure is low. If the light stays on for more than a few seconds after start-up, it's likely that the engine is worn and advice should be sought. If the light comes on while driving, **switch off** the engine immediately and seek advice.

IGNITION WARNING LIGHT

Acts as a reminder that the ignition circuit is switched on if the engine isn't running, and acts as a no-charge warning light. If the light stays on after starting, or comes on while driving, the battery is not charging properly.

The battery may therefore become fully discharged, and the best course of action is to stop and seek advice.

HEADLIGHT MAIN BEAM WARNING LIGHT

Acts as a reminder that the headlight main beam is switched on.

DIRECTION INDICATOR WARNING LIGHT

Shows that the direction indicators are switched on.

ABS (ANTI-LOCK BRAKING SYSTEM) WARNING LIGHT

This light should come on when the ignition is first switched on. If the light stays on for any longer than one minute after the engine has been started, or comes on while driving, there is a fault in one of the ABS circuits. Normal braking will still be available, but the braking system should be checked by a Ford dealer as soon as possible.

SEAT BELT WARNING LIGHT

This light comes on when the ignition is first switched on, and acts as a reminder that seat belts should be worn. The light should go out after approximately 20 seconds.

ECONOLIGHTS

These lights provide a rough indication as to how economically the car is being driven. Where fitted, the lights are positioned above the direction indicator warning light. As a guide, aim to drive so that the lights come on only momentarily, or not at all. The amber light will come on when the car is being driven less economically, and the red light will come on when the car is being driven uneconomically. In practice, it will be very difficult to drive without causing the lights to come on at some stage, especially when accelerating.

Auxiliary warning system

Consists of five lights which are connected to sensors to monitor various fluid levels and front brake pad wear. When the ignition is first switched on, the auxiliary warning lights will all come on for approximately 5 seconds. Check that the lights all go out before starting the engine. The function of each individual light is as follows.

BRAKE PAD WEAR WARNING LIGHT

Indicates that the front brake pads are worn to the stage where they require renewal. If the light comes on, the pads should be renewed as soon as possible.

LOW OIL LEVEL WARNING LIGHT

Indicates that the oil level has reached the 'MIN' mark on the dipstick. If the light comes on, the oil level should be topped up as soon as possible (refer to *'Regular checks'* on page 70). Note that the oil level check only works during initial start-up, and does not work with the engine running.

LOW COOLANT LEVEL WARNING LIGHT

Indicates that the coolant level has reached the 'MIN' mark on the reservoir. If the light comes on, the coolant level should be topped up as soon as possible (refer to *'Regular checks'* on page 70).

LOW FUEL LEVEL WARNING LIGHT

Comes on when there are approximately 10 litres (2.2 gallons) of fuel left in the tank.

LOW WASHER FLUID LEVEL WARNING LIGHT

Indicates that the windscreen washer fluid reservoir is below quarter-full.

Graphic warning system

Consists of a graphic display which uses symbols to warn of a light bulb failure, a door or tailgate/boot not properly closed, or a danger of ice. When the ignition is switched on, all the warning symbols should illuminate for approximately 5 seconds and then go out, with the exception of the brake light symbols. The brake light symbols should go out when the brake pedal has been pressed once. The symbols provide the following warnings:

DANGER OF ICE

A yellow symbol is displayed when the outside temperature falls to between 4 and 1°C. A red symbol is displayed when the temperature drops to 0°C or below.

BULB FAILURE

The paired light symbols will illuminate to indicate a faulty bulb or fuse.

DOOR/TAILGATE/BOOT NOT CLOSED

The relevant symbol will illuminate.

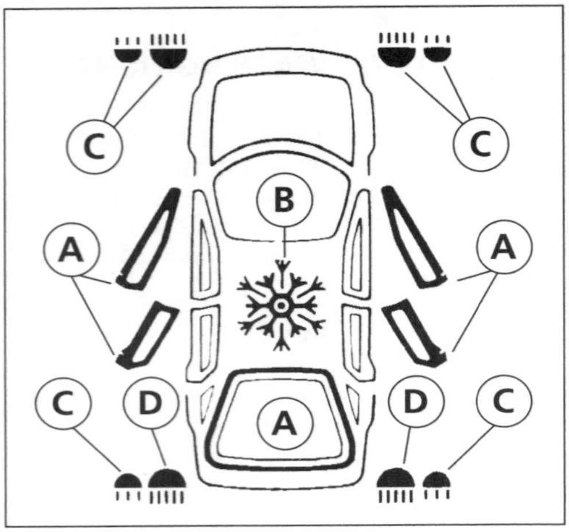

▲ *Graphic warning system display*

A Door or tailgate not closed
B Likelihood of frost or ice (yellow at 4°C, red at 0°C)
C Defective dipped headlight, sidelight, tail light, or related fuse or wiring
D Defective brake light, or related fuse or wiring

Ignition switch/steering lock

The switch has four positions as follows:

- **O** Ignition off, steering locked
- **I** Ignition off, accessory circuits on, steering unlocked
- **II** Ignition on, and all electrical circuits on
- **III** Starter motor operates (release the key immediately the engine starts)

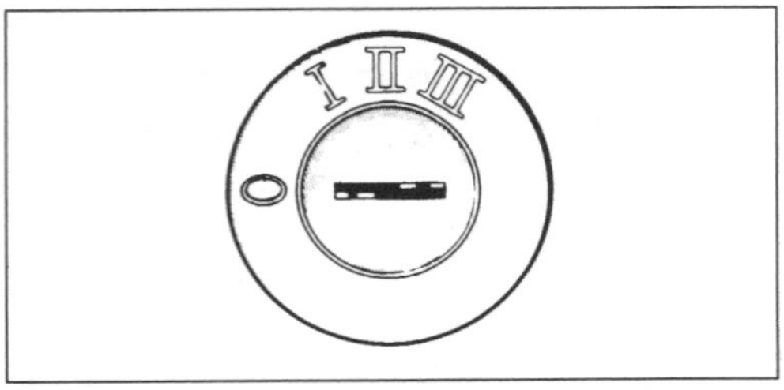

▲ *Ignition switch/steering lock positions (see text)*

Gearbox

MANUAL GEARBOX

Either a 4- or 5-speed gearbox may be fitted. The gear positions follow the usual 'H' pattern, but the method of selecting reverse gear varies depending on model. Reverse is selected in the same way for all types of gearbox except the 'MT 75' type 5-speed gearbox which was introduced in late 1988.

All gearboxes except 'MT 75' type

Reverse gear is positioned to the left of 1st gear, and is selected by moving the lever fully left from the neutral position, then pressing down against spring pressure and pushing forwards.

'MT 75' type gearbox

If an 'MT 75' type gearbox is fitted to your car, there will be a collar fitted under the gear lever knob.

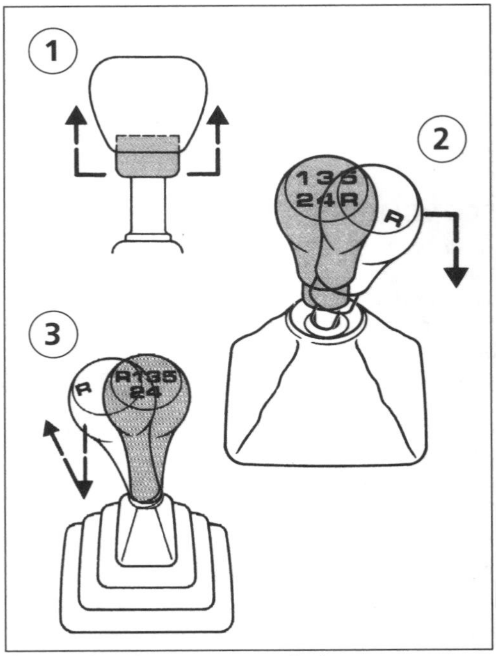

▲ *Manual gearbox reverse gear selection*

1 Collar under gear lever knob ('MT 75' type gearbox)
2 Selecting reverse ('MT 75' type gearbox)
3 Selecting reverse (All gearboxes except 'MT 75' type)

Reverse gear is positioned opposite 5th gear, and is selected by moving the lever fully right from the neutral position, then lifting the collar under the gear lever knob before pulling the lever rearwards.

AUTOMATIC TRANSMISSION

Two different types of automatic transmission have been fitted to Sierra models. Earlier models were fitted with a 3-speed transmission, but later models use a 4-speed unit. The 3-speed transmission has six selector positions, while the 4-speed unit has seven:

P – locks the transmission and should be used when parking. The engine can be started. Only select **P** after parking when the car is stationary, and always apply the handbrake.

R – selects **Reverse** gear.

N – selects **Neutral**. The engine can be started. No power is transmitted to the wheels.

D (or **DE** on some 4-speed transmissions) – selects **Drive**. The transmission will change gear automatically using 1st, 2nd and 3rd gears on 3-speed transmissions, or 1st, 2nd, 3rd and 4th gears on 4-speed transmissions.

3 (or **D** on some transmissions) – only available on 4-speed transmissions, allows the transmission to change automatically between 1st, 2nd and 3rd gears only.

2 – locks the transmission in 2nd gear. Useful when driving up or down hills, or driving through a series of bends. It should only be used at speeds above 25 mph (40 km/h).

1 – locks the transmission in 1st gear. It should only be used when driving up or down extremely steep slopes.

The handbrake or footbrake must be applied before selecting **R**, **DE**, **D**, **3** or **1**, when the car is stationary. To select positions **R** and **P**, the selector lever must first be pushed down in the **N** position. The lever must also be pushed down to move it out of the **P** position.

Multi-function switch

Controls the direction indicators, headlight flash and main beam, and on certain models the horn (if there is no horn button on the multi-function switch, the horn is controlled by a button in the centre of the steering wheel). On models fitted with driving lights, the driving lights operate with the headlight main beam and headlight flash.

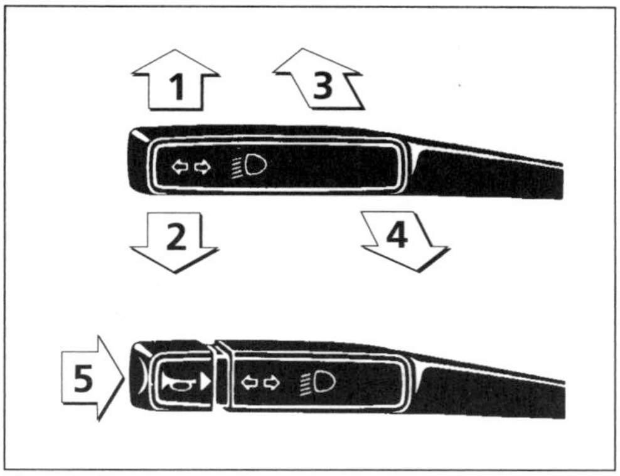

▲ *Multi-function switch*
1 Right direction indicator
2 Left direction indicator
3 Headlight main beam (and driving lights where fitted)
4 Headlight flash
5 Horn (certain models only)

Exterior lights switch

Controls the sidelights/tail lights and headlights.

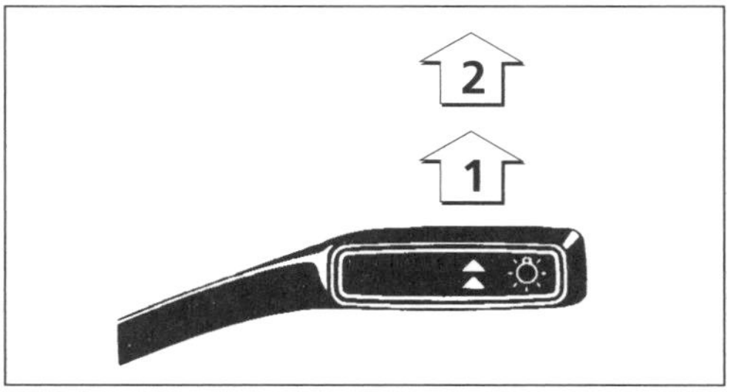

▲ *Exterior lights switch*
1 Sidelights/tail lights
2 Headlights

Windscreen wash/wipe switch (models up to 1987)

Controls the windscreen wipers and washers. Where fitted, the headlight washers will operate at the same time as the windscreen washers when the headlights are switched on. On certain models with intermittent wipe, the interval between wipes can be varied using the

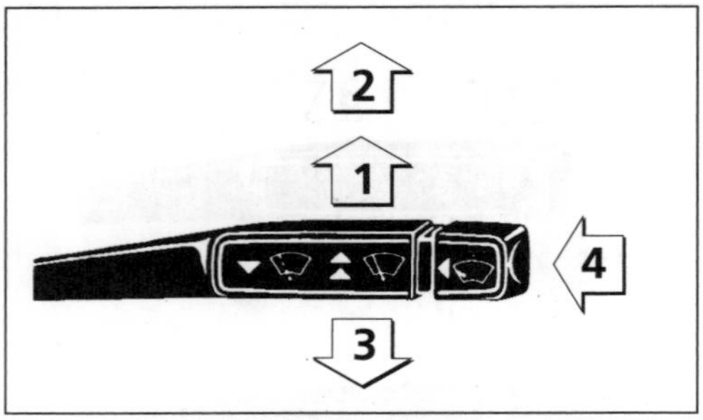

▲ *Windscreen wash/wipe switch (models up to 1987)*

- ***1*** *Normal wiper speed*
- ***2*** *Fast wipe*
- ***3*** *Single wipe or intermittent wipe (depending on model)*
- ***4*** *Windscreen washers*

wheel provided on the facia. Turn the wheel right to increase the interval, or left to decrease the interval.

Tailgate wash/wipe switch (models up to 1987)

The switch is positioned on the right-hand side of the facia. Depress the switch to the first position to operate the wiper. Depress the switch further and hold it depressed to operate the washer.

Windscreen and tailgate (where applicable) wash/wipe switch (models from 1987)

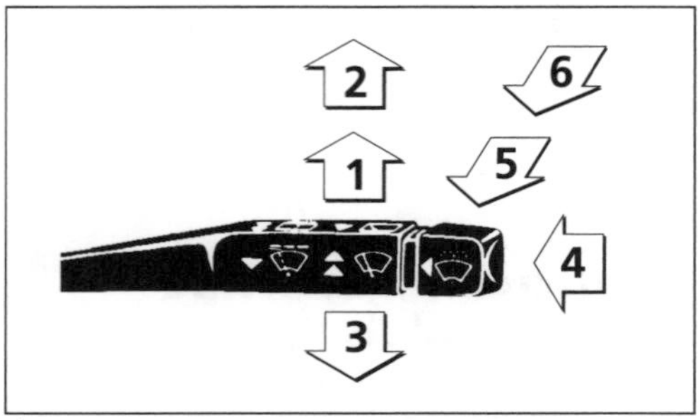

▲ *Windscreen and tailgate wash/wipe switch (models from 1987)*

- ***1*** *Normal wiper speed*
- ***2*** *Fast wipe*
- ***3*** *Intermittent wipe*
- ***4*** *Windscreen washers*
- ***5*** *Tailgate wiper*
- ***6*** *Tailgate washer*

Controls all the wash/wipe functions. On certain models with intermittent wipe, the interval between wipes can be varied using the wheel provided on the facia. Turn the wheel right to increase the interval, or left to decrease the interval.

Heating and ventilation controls

Lever **(A)** – controls the temperature: up for hot, and down for cold, with variable settings in between.

Lever **(B)** – controls the distribution of the air between the windscreen (up) and the car floor area (down), with variable settings in between. When the lever is fully down, the airflow is shut off.

Switch **(C)** – controls the booster fan. The fan has three speeds.

The airflow through the louvred vents in the facia is controlled independently using the knurled wheel at the side of each vent to increase or decrease the amount of air, and by moving the position of the slats to control the direction.

Side vents at each end of the facia operate with the windscreen demister vents to de-mist the side windows.

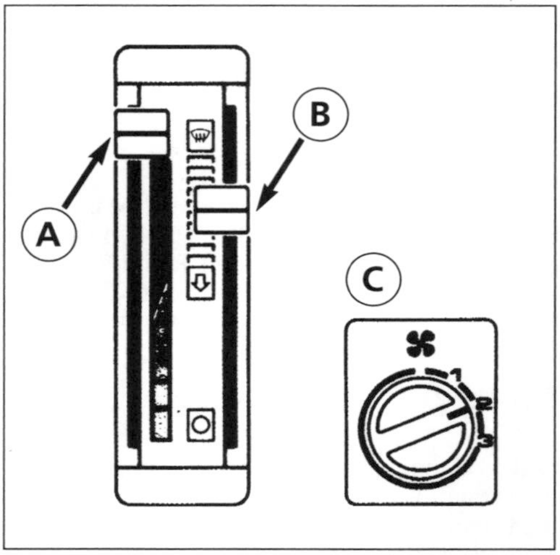

▲ *Heating and ventilation controls*

- ***A*** *Air temperature control*
- ***B*** *Air distribution control*
- ***C*** *Booster fan switch*

Air conditioning controls

Lever **(A)** – is described under 'Heating and ventilation controls'.

Lever **(B)** – is described under 'Heating and ventilation controls'.

Switch **(C)** – controls the air conditioning system. A warning light **(E)** will illuminate when the system is switched on. To switch on the air conditioning system, push in the knob, and turn it clockwise to select fresh air, or anti-clockwise to select recirculated air. In both cases the booster fan has three speeds. Use levers **(A)** and **(B)** to adjust the airflow and temperature.

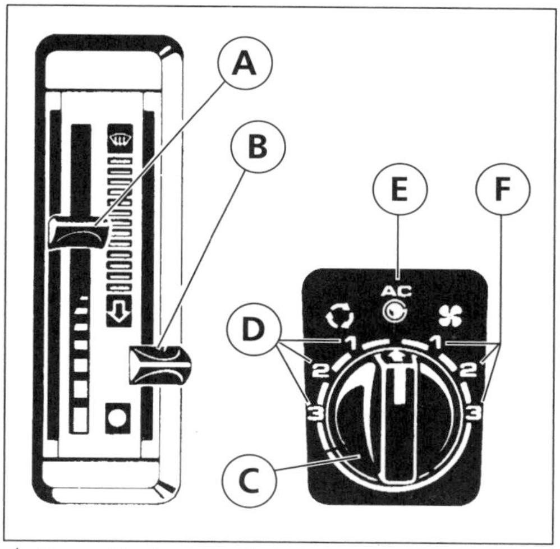

▲ *Air conditioning controls*
A Air temperature control
B Air distribution control
C Air conditioning control switch
D Recirculated air
E Warning light
F Fresh air

Hazard flasher switch

Operates the flashers regardless of the position of the ignition switch.

Rear foglight switch

Operates the rear foglights, provided that the headlights are switched on at dipped beam.

Front foglight switch

Operates the front foglights, provided that the sidelights or headlights are switched on.

Heated rear window switch

Operates the heated rear window, provided that the ignition is switched on. To avoid battery drain, it's recommended that the device is switched off as soon as demisting is complete (some models are fitted with a timer which switches the element off automatically after 10 minutes). On models fitted with electrically-adjustable mirrors, the mirror heating elements will operate when the heated rear window is switched on.

Electric door mirror switch (models up to 1987)

Located in the driver's door panel. To adjust the driver's mirror, move switch **(1)** to the right, then move switch **(2)** in the required direction. To adjust the passenger's mirror, move switch **(1)** to the left. The mirror heating elements will operate when the heated rear window is switched on.

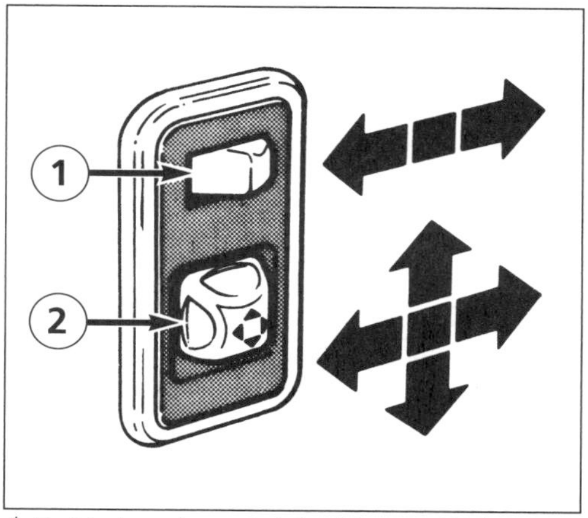

▲ *Electric door mirror switch (models up to 1987)*
*For **1** and **2** see text*

Electric door mirror switch (models from 1987)

The switch has three positions to adjust the driver's and passenger's door mirrors:

A – the passenger's door mirror can be adjusted by moving the knob in the required direction.

B – off.

C – the driver's door mirror can be adjusted by moving the knob in the required direction.

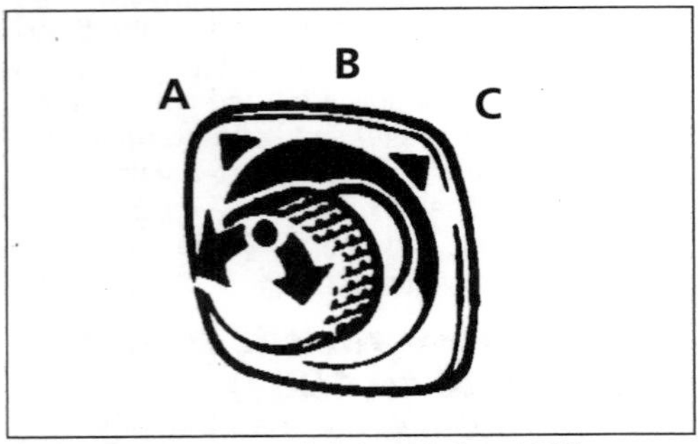

▲ *Electric door mirror switch (models from 1987) For **A**, **B** and **C** see text*

Heated windscreen switch

Operates the windscreen heating element provided that the engine is running. A timer is fitted which will switch the element off automatically after 4 minutes.

Heated front seat switch(es)

Mounted in the facia, or in the centre console. The switch(es) operate(s) the heating element(s) with the ignition key in position **I** or **II**. To avoid battery drain, the heated seat(s) should only be used with the engine running.

Loudspeaker balance control

Adjusts the sound balance between the front, rear, left, and right of the car by moving the joystick in the required direction.

Instrument illumination control

Adjusts the brightness of the instrument illumination when the exterior lights are switched on.

Remote tailgate release (early models)

Operates with the ignition key in position **I** or **II**, provided that the engine is not running.

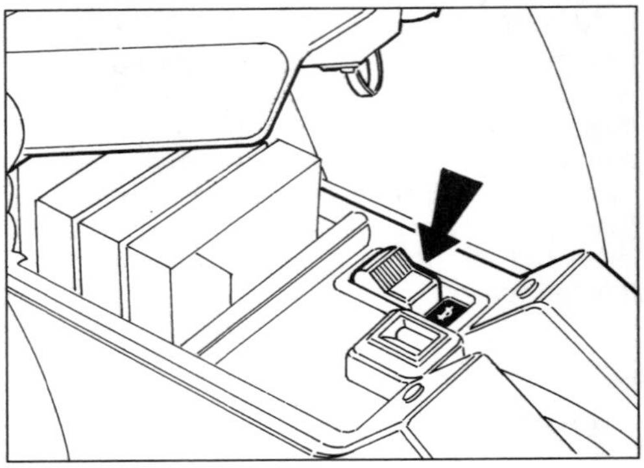

▲ *Remote tailgate release (early models)*

Remote tailgate/boot and fuel filler flap release

Mounted to the right of the driver's seat. Pull the lever up to open the tailgate/boot, and push the lever down to open the fuel filler flap.

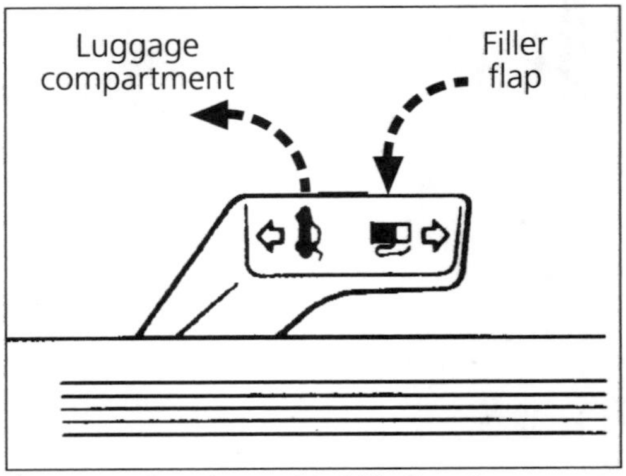

▲ *Remote tailgate/boot and fuel filler flap release*

Bonnet release lever

Mounted under the steering column. Pull the lever to release the bonnet, then lift the bonnet safety catch (slightly above and right of Ford badge) to open the bonnet.

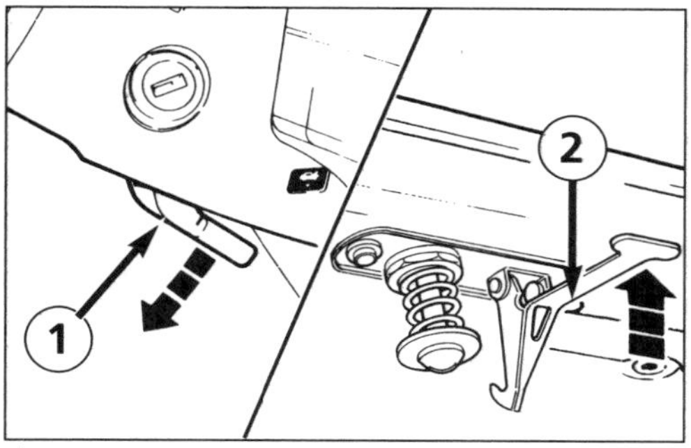

▲ *Bonnet release mechanism*
1 Bonnet release lever
2 Safety catch

INTERIOR EQUIPMENT

Cigarette lighter

Operates regardless of the position of the ignition switch. Press the lighter in, then release it, and wait until it pops out ready for use.

Electric windows

Switches mounted in the centre console raise and lower the front, and on some models the rear, windows with the ignition switched on. On models with rear electric windows, additional switches for the rear seat passengers may be located in the rear of the centre console, or in the rear door armrests. Press the rear of the relevant switch to lower (open) the window, or the front of the switch to raise (close) the window. If rear electric windows are fitted, a cut-out switch is provided between the other window switches in the front of the centre console. In its raised position, the cut-out switch will isolate the electric window switches in the rear of the car, to prevent the windows being operated by the rear seat passengers. This can be useful when carrying children.

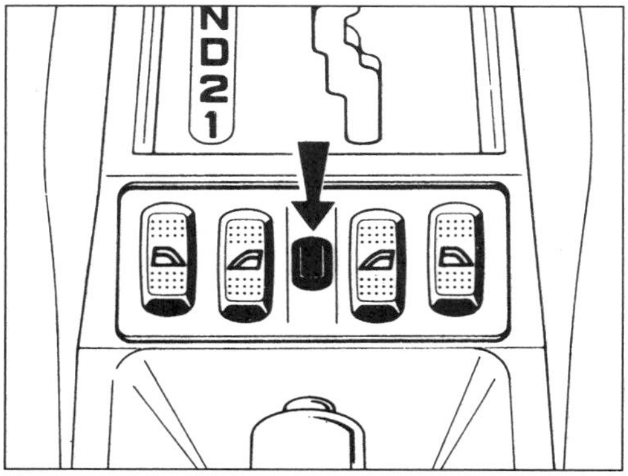

▲ *Rear electric windows cut-out switch (arrowed)*

Sunroof

The sunroof is of the tilt/sliding type. To open the sunroof, ease the winder handle out of its recess, and turn it anti-clockwise. Turn the handle clockwise to close the sunroof. To tilt the rear edge of the sunroof, ease the winder handle out of its recess, and turn it clockwise against a slight resistance until the sunroof tilts. Open the facia side air vents to prevent a partial vacuum inside the car with the roof in this position. The sun-blind moves back when the sunroof is opened, and can be moved to any position when the sunroof is closed or tilted.

Dipping rear view mirror

Pull back the lever under the mirror to reduce the glare from the lights of following vehicles.

Seat belts

The operation of the seat belts is self-explanatory, but certain points require additional explanation.

FRONT SEAT BELT HEIGHT ADJUSTMENT

On certain models, the height of the upper seat belt mounting can be adjusted by pushing the button in the centre of the adjuster and sliding the adjuster up or down as required. Make sure that the adjuster engages in one of the preset positions with a positive click.

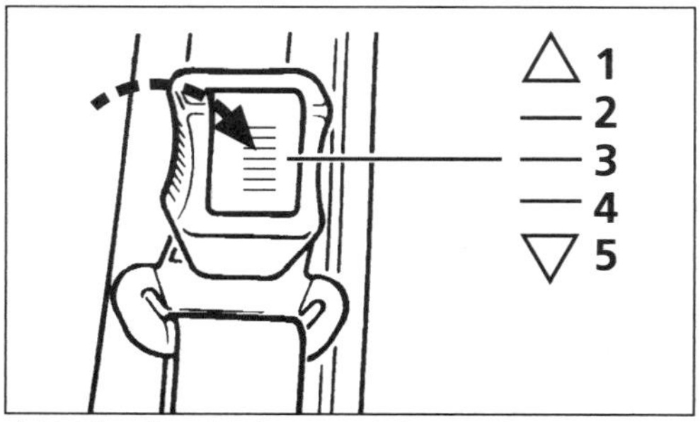

▲ *Height-adjustable front seat belt mounting. Press button (arrowed) to adjust*

REAR SEAT BELTS

The centre rear seat belt is of the lap type, and must be manually adjusted for fit by the wearer. To tighten the belt, pull the webbing through the adjuster, and move the plastic clip to hold the end of the belt. To lengthen the belt, turn the adjuster at right-angles to the belt, then press the tongue into the adjuster housing and pull the webbing through the adjuster.

Front seat adjustment

SEAT SLIDING CONTROL

To move a seat backwards or forwards, pull the lever located under the front of the seat away from the door on early models, or up towards the seat on later models; then slide the seat to the required position. Release the lever, and ensure that the seat has locked in position by gently rocking it backwards and forwards.

SEAT BACK RELEASE (3-DOOR MODELS)

Pull up the lever on the side of the seat, and tilt the seat back to allow access to the rear seats.

SEAT RECLINING CONTROL

Turn the wheel on the outside bottom edge of the seat to alter the angle of the seat back.

SEAT LUMBAR ADJUSTMENT

On certain models, the seat lumbar support can be adjusted, using either a wheel on the inside edge of the seat back, or by using a hand pump, depending on model. On models fitted with a hand pump, the pump is located on the lower inside edge of the seat. To increase the support, squeeze the inflator **(A)**. To reduce the support, push the button **(B)**. Push the inflator down beside the seat when not in use.

DRIVER'S SEAT HEIGHT ADJUSTMENT

On certain models, the height of the driver's seat can be adjusted by turning the handle located at the front lower edge of the seat. To raise the seat turn the handle anti-clockwise, and to lower the seat turn the handle clockwise.

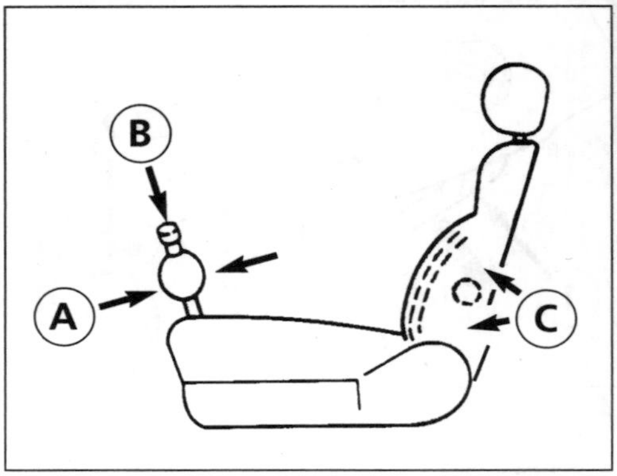

▲ *Front seat lumbar adjustment*
 A *Inflator (to increase support)*
 B *Button (to reduce support)*
 C *Adjuster wheel (alternative to hand pump*

Adjustable head restraints

The head restraints can be adjusted for height by pulling up or down (do not force the head restraint beyond the upper stop position), and the angle can be adjusted by tilting backwards or forwards.

Folding rear seats

Either a full-width or a split rear seat back may be fitted, depending on model. Make sure that the front seats are positioned to allow adequate clearance for the rear seats to be folded forwards.

SALOON (SAPPHIRE) MODELS

The release catches for the seat back(s) are only accessible from the luggage compartment. Open the boot and pull the release catch(es), then fold the seat back(s) forwards.

When returning the seat back(s) to the normal position, hold the top(s) of the rear seat belt(s) forwards to avoid trapping, and make sure that the seat back locking catch(es) engage properly.

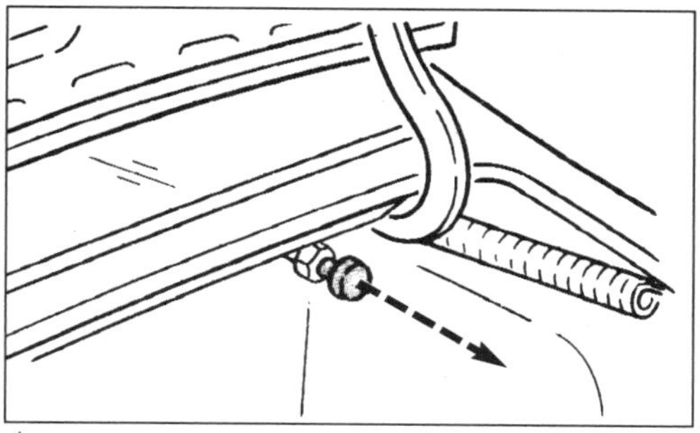

▲ *Rear seat back release catch (Saloon models)*

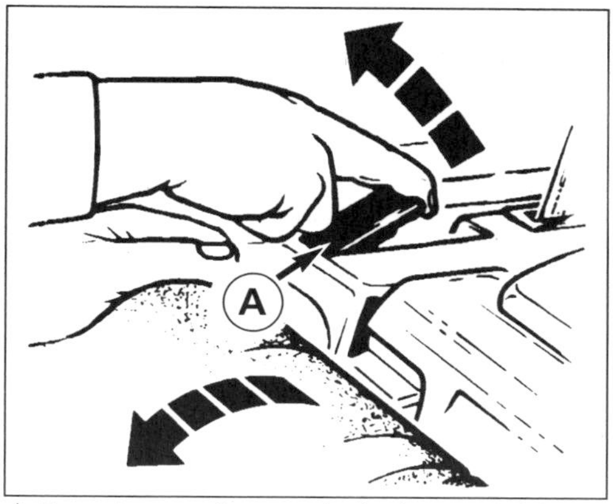

▲ *Alternative type of rear seat back release catch* ***(A)*** *(Hatchback and Estate models)*

HATCHBACK MODELS

Maximum load-carrying capacity is obtained by folding the rear seat cushion forwards against the front seats before folding down the rear seat back(s). If desired, pull the cushion forward using the looped tabs provided.

Early models are fitted with safety catches which must be released before the seat back(s) can be folded. The safety catches are accessible from the luggage compartment, and must be pushed towards the centre of the car to release the locking mechanism.

With the safety catches released (where applicable), pull the release catch(es) on the top of the seat back(s), and fold the seat back(s) forward.

When returning the seat back(s) to the normal position, hold the top(s) of the rear seat belt(s) forward to avoid trapping, and make sure that the seat back locking catch(es) engage properly (where safety catches are fitted, push the catch(es) towards the side(s) of the car to lock them). Where applicable, also make sure that the seat cushion locks into position, and that the seat belts are not trapped.

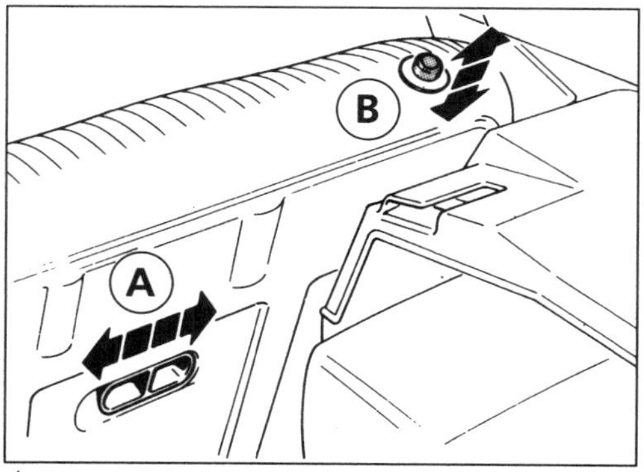

▲ *Rear seat back safety catch* ***(A)*** *and release catch* ***(B)*** *(early Hatchback models)*

ESTATE MODELS

The release catches for the seat back(s) are located on the top edges of the seat back(s). Pull the release catch(es) and then fold the seat back(s) forward. Some early models are fitted with safety catches to hold the seat back(s) in the lowered position – make sure that the catch(es) is/are engaged where applicable.

▲ *Rear seat back safety catch (early Estate models)*

To return the seat back(s) to the normal position, on models with safety catches, push the catch lever(s) down to release the seat back(s). When returning the seat back(s) to the normal position, hold the top(s) of the rear seat belt(s) forwards to avoid trapping, and make sure that the seat back locking catch(es) engage properly.

Rear parcel shelf

On Hatchback and Estate models with a rear parcel shelf, the shelf can be removed by unhooking the cords from the tailgate, and pulling the shelf rearwards and upwards.

Rear window sun-blinds

Certain models are fitted with a two-piece sun-blind on the inside of the rear window. To operate a blind, pull upwards and fasten the ends of the blind to the hook provided on the rear window frame. The blinds retract automatically when unhooked.

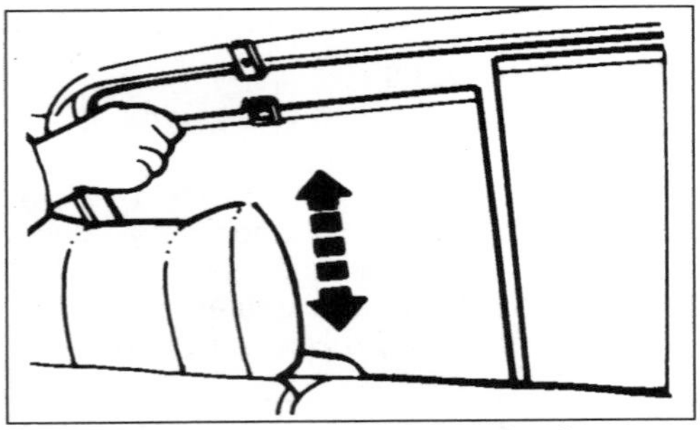

▲ *Rear window sun-blinds hook onto window frame*

EXTERIOR EQUIPMENT

Door locks

To lock a front door, either close it and turn the key in the lock, or press the rocker switch located in the door interior handle housing, so that the red warning indicator is *not* visible. This can be done with the doors open, in which case the doors will be locked when they are shut. Note that on later models, the driver's door can only be locked using the key. The front doors can be opened from inside with the rocker switches in the 'locked' position using the interior handle in the usual way.

To lock rear door, press the rocker switch located in the door interior handle housing, so that the red warning indicator is *not* visible. This can be done with the doors open, in which case the doors will be locked when they are shut. To open a rear door from the inside, the rocker switch must be in the 'unlocked' position. The rear door locks incorporate childproof catches, operated by pushing the small lever on the back edge of the door downwards on early models, or towards the interior of the car on later models.

On models fitted with an anti-theft alarm system, the system will be activated as soon as the driver's or front passenger's door is locked with the key.

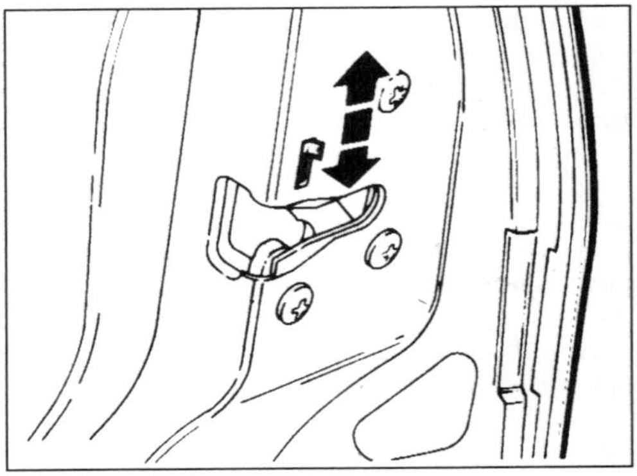

▲ *Childproof catch operating lever (early models)*

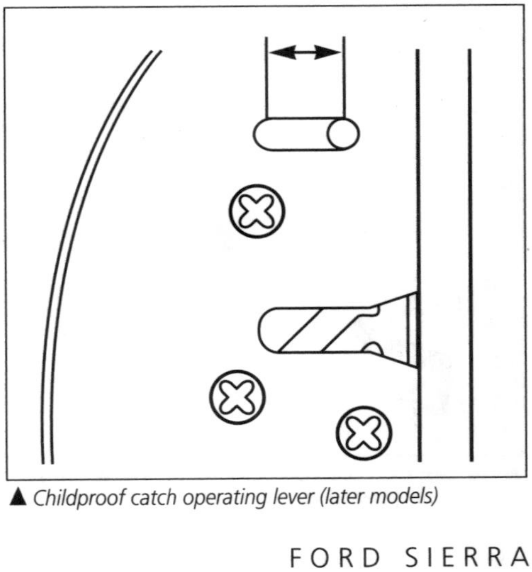

▲ *Childproof catch operating lever (later models)*

Tailgate/boot lid lock

To unlock the tailgate/boot lid, turn the key clockwise. The lock button must be pushed in to open the tailgate/boot lid on early models.

Central door locking system

The system can be operated from the driver's door on models up to 1987, or from the driver's and passenger's door on models from 1987. On the earlier models, the front passenger's door lock and the tailgate/boot lid lock can be operated independently of the driver's door. On the later models, the tailgate/boot lid cannot be opened with the engine running. It's only possible to operate the system with the doors closed, either by locking from outside using the key, or from inside using the lock rocker switch.

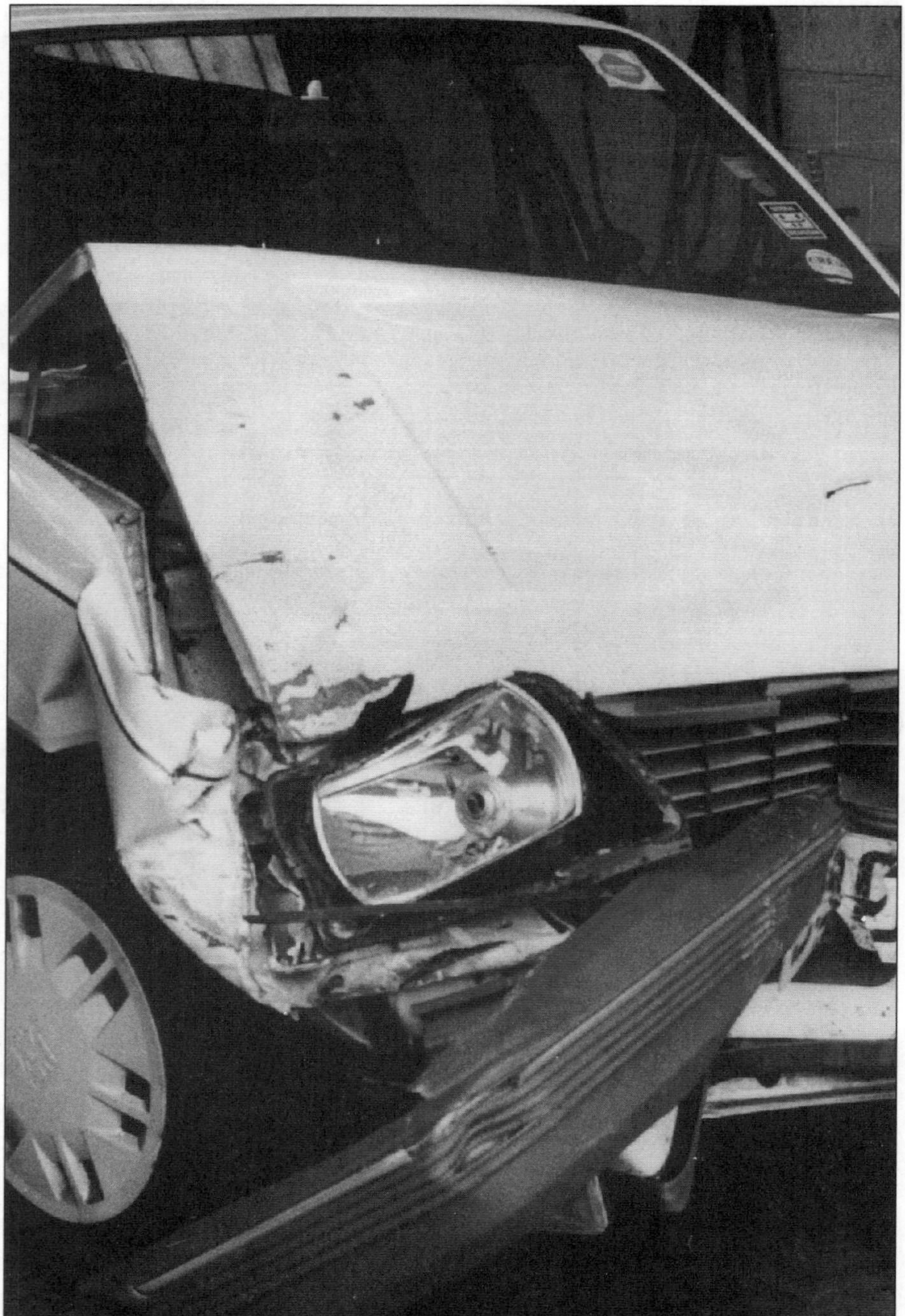

FORD SIERRA

In the event of an accident, the first priority is safety. This may seem obvious, but in the heat of the moment, it's very easy to overlook certain points which may worsen the situation, or even cause another accident. The course of action to be taken will vary depending on how serious the accident is, and whether anyone is injured, but always try to think clearly, and don't panic.

We don't suggest that you consult this Section at the scene of an accident, but hopefully the following advice will help you to be better prepared to deal with the situation should you be unfortunate enough to appear on the scene of an accident or become involved in one yourself.

HOW TO COPE WITH AN ACCIDENT

Deal with any possible further danger

Further collisions and fire are the dangers in a road accident.

- **If possible warn other traffic**

  Where possible, switch on the car's hazard warning flashers.
  If a warning triangle is carried, position it a reasonable distance away from the scene of the accident, to give approaching drivers sufficient warning to enable them to slow down. Decide from which direction the approaching traffic will have least warning, and position the triangle accordingly.
  If possible, send someone to warn approaching traffic of the danger which exists, and to encourage the traffic to slow down.

- **Switch off the ignition and impose a 'No Smoking' ban**

  This will reduce the possibility of fire, should there be a petrol leak.

Call for assistance

Send someone to call the emergency services (dial 999), and make sure that all the necessary information is provided to the operator. Give the exact location of the accident, and the number of vehicles and if applicable the number of casualties involved. Refer to the *'Motorway breakdowns'* Section on page 44 for details of how to call for assistance on a motorway.

- **Call an ambulance**

  If anyone is seriously injured or trapped in a vehicle.

- **Call the fire brigade**

  If anyone is trapped in a vehicle, or if you think that there is a risk of fire.

- **Call the police**

  If any of the above conditions apply, or if the accident is a hazard to other traffic. In most cases, the accident must be reported to the police within 24 hours even if none of the above conditions apply (refer to *'Requirements of the law in the event of an accident'* on page 38).

Administer first aid

Refer to *'First aid'* on page 38.

Provide your details

If you are involved in the accident as a driver or car owner, provide your personal and vehicle details to anyone having reasonable grounds to ask for them. Also inform the police of the details of the accident as soon as possible, if not already done.

FIRST AID

Always carry a first aid kit

If possible, learn first aid by attending a suitable course – contact the St John Ambulance Association or Brigade, St Andrew's Ambulance Association or the British Red Cross Society in your local area for details. These organisations will also be able to provide further training in heart massage and mouth-to-mouth resuscitation which could enable you to save someone's life in an emergency.

Before proceeding with any kind of first aid treatment, deal with any possible further danger, and call for assistance, as described previously in this Chapter.

To help remember the sequence of action which should be followed when dealing with a seriously injured casualty, use the **ABC** of emergency first aid.

Casualties remaining in vehicles

Any injured people remaining in vehicles should not be moved unless there is a risk of further danger (such as from fire, further collisions etc).

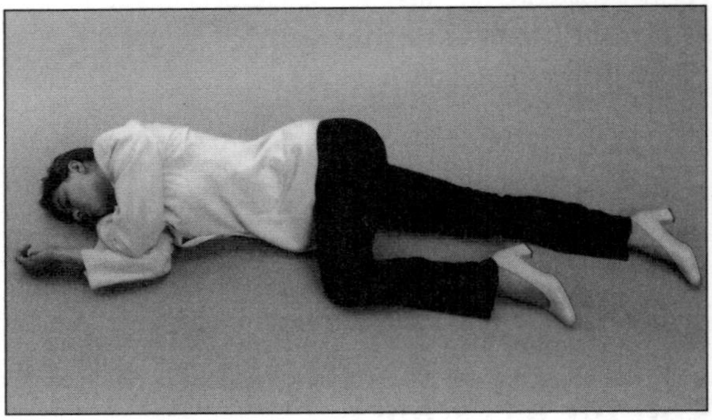

▲ *The recovery position*

Casualties outside the vehicles

Recovery position

To prevent the possibility of an unconscious but breathing casualty choking or suffocating, he/she must be placed in the recovery position.

Lie the casualty on his/her side.

Bend the leg and position the arms as shown to support the head.

First aid for other injuries

Always treat injuries in the following order of priority:

- Ensure that the casualty has a clear airway. If not treat as described opposite.
- Put any breathing, unconscious casualty into the recovery position as described previously.
- Treat any severe bleeding by applying direct pressure to the wound. Maintain the pressure for at least 10 minutes and apply a suitable clean dressing. If the wound continues to bleed, apply further pressure and dressings **over** that already in place. If bleeding from a limb, and as long as the limb is not broken, lift the affected limb to reduce the bleeding.
- Broken limbs should not be moved, unless the casualty has to be moved in order to avoid further injury. Support the affected limb(s) by placing blankets, bags, etc, alongside.
- Burns should be treated with plenty of cold water as soon as possible. Don't attempt to remove any clothing, but cover with a clean dressing.

Reassurance

The casualty may be in shock, but prompt treatment will minimise this. Reassure the casualty confidently, avoid unnecessary movement, and keep him/her comfortable and warm. Make sure that the casualty is not left alone.

Give the casualty NOTHING to eat, drink or smoke.

A

Airway

- **Check for breathing**

Check that the airway is clear. If the casualty is breathing noisily or appears not to be breathing at all, remove any obvious obstruction in the mouth and tilt the head back as far as possible. Breathing may then start.

B

Breathing

- **If breathing has stopped**

If after clearing the airway the casualty still appears not to be breathing, look to see if the chest or abdomen is moving. Place your ear close to the casualty's mouth to listen and feel for breathing, and look to see if there's any movement of the chest. If you can't detect anything, you must start breathing for the casualty (mouth-to-mouth resuscitation).

To do this, pinch the casualty's nostrils firmly, keep his/her chin raised, and seal your lips around the casualty's mouth. Breathe out through your mouth into the casualty until the chest rises, then remove your mouth and allow the casualty's chest to fall. Give one further breath.

If giving mouth-to-mouth resuscitation to a child, remember that an adult's lungs are significantly larger than those of a child. Care must be taken not to overinflate a child's lungs, as this may cause injury.

- **Check for a pulse at the base of the neck**

If the heart is beating, continue to breathe into the casualty at a rate of one breath every 5 seconds, until he/she is able to breathe unaided, then place the casualty in the recovery position (refer to *'Recovery position'* opposite).

C

Circulation

- **If the heart has stopped**

If there is no pulse at the neck, then external chest compression (heart massage) must be started.

To do this, if the casualty is still in the vehicle, he/she must be removed and laid on his/her back on the ground. Kneel down on one side of the casualty. Feel for the lower half of the casualty's breastbone and place the heel of one of your hands on this part of the bone, keeping your fingers off the casualty's chest. Cover this hand with your other hand, interlocking your fingers. Keeping your arms straight and vertical, press the breastbone down about 4 to 5 centimetres (1½ to 2 inches). The pressure should be smooth but not jerky. Do this 15 times, at the rate of just over once a second. It may help you to count aloud as you press.

Follow this by 2 breaths as described under *'Breathing'*.

Repeat the cycle of 2 breaths followed by 15 compressions, twice more, ending with 2 breaths, then re-check the pulse.

- **If there is still no pulse**

Repeat the cycle 9 times and then check the pulse. Continue this procedure until there is a pulse or until professional help arrives.

REQUIREMENTS OF THE LAW IN THE EVENT OF AN ACCIDENT

The following is taken from the Road Traffic Act of 1988.

If you are involved in an accident – which causes damage or injury to any other person, or another vehicle, or any animal (horse, cattle, ass, mule, sheep, pig, goat or dog) not in your vehicle, or roadside property:

You must

- stop;
- give your own and the vehicle owner's name and address and the registration mark of the vehicle to anyone having reasonable grounds for requiring them;
- if you do not give your name and address to any such person at the time, report the accident to the police as soon as reasonably practicable, and in any case within 24 hours;
- if anyone is injured and you do not produce your certificate of insurance at the time to the police or to anyone who has with reasonable grounds required its production, report the accident to the police as soon as possible, and in any case within 24 hours, and either produce your certificate of insurance to the police when reporting the accident or ensure that it is produced within seven days thereafter at any police station you select.

ESSENTIAL DETAILS TO RECORD AFTER AN ACCIDENT

If you're involved in an accident, note down the following details which will help you to complete the accident report form for your insurance company, and will help you if the police become involved.

- The name and address of the other driver and those of the vehicle owner, if different.
- The name(s) and address(es) of any witness(es) (independent witnesses are particularly important).
- A description of any injury to yourself or others.
- Details of any damage caused to the vehicles involved or other property.
- The name and address of the other driver's insurance company and, if possible, the number of his/her certificate of insurance.
- The registration number of the other vehicle (check this against the tax disc if possible).
- The number of any police officer attending the scene.
- The location, time and date of the accident.
- The speed of the vehicles involved.
- The width of the road, details of road markings and signs, the state of the road surface, and the weather conditions.
- Any marks or debris on the road relevant to the accident.
- A rough sketch showing the vehicle positions before and after the accident. It's helpful to make a note of the vehicle positions in terms of distance from fixed landmarks, such as lamp posts, buildings, etc.
- Whether any of the other vehicle occupants were wearing seat belts.
- If the accident occurred at night or in poor visibility, whether vehicle lights or street lights were switched on.
- If you have a camera, take a picture of the vehicles and the scene.

If the other driver refuses to give you his/her name and address, or if you consider that he/she has committed a criminal offence, inform the police immediately.

Use this page to record all the relevant details in the event of an accident.
You should transfer all the information recorded on this page onto the Motor Vehicle Accident Report form which you can obtain from your insurance company.

OTHER DRIVER DETAILS

FULL NAME OF DRIVER:

ADDRESS:

POST CODE:

HOME TELEPHONE: WORK TELEPHONE:

DRIVING LICENCE NUMBER: ISSUED:

DATE OF BIRTH: DATE DRIVING TEST PASSED:

TYPE OF LICENCE: (tick box) Full ☐ Provisional ☐ Heavy Goods ☐

PERMITTED GROUPS:

FULL NAME OF OWNER:

ADDRESS:

POST CODE:

TELEPHONE:

INSURANCE COMPANY:

POLICY NUMBER:

AGENT OR BROKER:

TELEPHONE:

OTHER VEHICLE DETAILS

MAKE: MODEL:

YEAR: CC:

REGISTRATION NUMBER:

ACCIDENT REPORT DETAILS

CIRCUMSTANCES OF ACCIDENT

DATE: TIME: am/pm

PLACE: (Street or Road)

TOWN: COUNTY:

SPEED: WERE THE POLICE CALLED? Yes ☐ No ☐

If 'Yes' give details of Police Constabulary concerned:

DETAILS OF WHAT HAPPENED (Please use the page opposite to make a rough sketch map)

WHAT WAS THE WEATHER LIKE?

INDEPENDENT WITNESS 1

TELEPHONE:

INDEPENDENT WITNESS 2

TELEPHONE:

DETAILS OF ANY INJURIES SUSTAINED BY EITHER PARTY:

ACCIDENT REPORT SKETCH MAP

Essential items to include in your sketch map:

- The layout of the road and its approaches.
- The directions and identities of both vehicles.
- Their relative positions at the time of impact.
- The road signs and road markings.
- Names of the Streets and Roads.

HOW TO COPE WITH A FIRE

Fire is unpredictable, and particularly dangerous when cars are involved because of the presence of petrol, which is highly inflammable.

It's sensible to carry a fire extinguisher on board your car, but bear in mind that the average car fire extinguisher is suitable for use only on the smallest of fires.

In the event of your car catching fire:

- **Switch off the ignition.**
- **Get any passengers and yourself out of the car and well away from danger.**
- **Impose an immediate 'No Smoking' ban.**
- **DO NOT expose yourself, or anyone else, to unnecessary risk in an attempt to control the fire.** Minor fires can be controlled using a suitable extinguisher, but always use an extinguisher at arms-length, and bear in mind that a small vehicle fire can develop into a serious situation without warning.
- **Call for assistance, if necessary** - refer to *'How to cope with an accident'* on page 35.

HOW TO COPE WITH A BROKEN WINDSCREEN

There are a number of national specialist companies who offer roadside assistance to drivers who suffer broken windscreens, and it's a good idea to carry the 'phone number of a suitable specialist in your car for use in such situations. Some car insurance policies enable the use of such companies at preferential rates, and it's worth enquiring about this when obtaining insurance quotes.

- **Most of the windscreens fitted to modern cars are of the laminated type.** Laminated windscreens are made from layers of glass and plastic (usually two layers of glass sandwiching a layer of plastic) which prevents the windscreen from shattering, and preserves the driver's vision in the event of the screen suffering an impact.

  A sharp impact may crack the outer glass layer, but clear vision is usually maintained.

  Obviously, if the windscreen is damaged, it should be replaced at the earliest opportunity, but there is no need to curtail your journey unless the damage is particularly serious.
- **Some older cars may be fitted with toughened windscreens.** If a toughened windscreen suffers an impact, the glass will normally stay intact, but severe 'crazing' will usually occur which is likely to seriously affect the driver's vision.

  If a toughened windscreen breaks, stop immediately, and seek assistance. **DO NOT** attempt to knock the broken glass out of the windscreen frame, as this is likely to result in injury to yourself, and damage to the car. Driving without a windscreen is extremely dangerous and should not be attempted.

WHAT TO DO IF YOUR CAR IS BROKEN INTO OR VANDALISED

If you are unfortunate enough to have your car broken into or vandalised, do not attempt to drive your car away from the scene unless you are satisfied that no damage has been caused which could affect the car's safety.

Where possible notify the police before moving your car, and inform your insurance company at the earliest opportunity.

In the unfortunate event of a breakdown, the first priority must always be safety – never risk causing an accident by attempting to move the car or trying to work on it if it's in a dangerous position.

Joining one of the national motoring organisations such as the AA or the RAC can provide you with a recovery and assistance service should you break down. It may also save you money, as local garages often charge high rates to provide a recovery service.

This Section provides advice on the course of action to follow should you break down. There are also Sections covering the procedure to follow for towing, and on how to cope with two of the more common causes of breakdowns – a puncture, and a flat battery.

Breakdowns on an ordinary road (not a motorway)

Most importantly of all, common sense must be used, but the following advice should prove helpful.

- **Warn approaching traffic,** to minimise the risk of a collision. Where possible, switch on the car's hazard warning flashers, and at night leave the sidelights switched on.
  If a warning triangle is carried, position it a reasonable distance away from the scene of the breakdown, to give approaching drivers sufficient warning to enable them to slow down. Decide from which direction the approaching traffic will have least warning, and position the triangle accordingly. Place the warning triangle on the road surface, out from the edge of the road, where it can be easily seen.
  If possible, send someone to warn approaching traffic of the danger which exists, and to encourage the traffic to slow down.
- **Get the passengers out of the car** as a precaution should the car be hit by another vehicle. The passengers should move well away from the car and approaching traffic, up (or down) an embankment, or into a nearby field for example.
- **If possible, move the car to a safe place.** If the car cannot be safely driven or pushed to a place of safety, call a recovery service to provide assistance. *Don't expose yourself or anyone else to unnecessary danger in order to move the car.*
- **If the car can't be safely moved, turn the steering wheel towards the side of the road.** If the car is hit it will then be pushed into the side of the road, not into the path of traffic.
- **Try to find the cause of the problem.** Only attempt this if the car is in a safe position away from traffic. Refer to *'Fault finding'* on page 131 for details.
- **If necessary call for assistance** from one of the national motoring organisations if you're a member, or from a local garage who can provide a recovery service.

Motorway breakdowns

In the event of a motorway breakdown, due to the speed and amount of traffic there are a few special points to consider, and the following advice should be followed.

- **If possible, switch on the car's hazard warning flashers, and at night leave the sidelights switched on.**
- **DO NOT open the doors nearest to the carriageway, and DO NOT stand at the rear of the car, or between it and the passing traffic.**
- **Get the car off the carriageway and onto the hard shoulder as quickly as possible, and as far to the left as possible** – never forget the danger from passing traffic. Turn the steering wheel to the left so that the car will be pushed away from the carriageway if hit from behind.
- **If you can't move the car off the carriageway** – get everyone out of the car and well away from the carriageways (up/down an embankment, or into a nearby field for example) as quickly as possible. Never forget the danger from passing traffic.

  Call for assistance as described below. Attempt to warn approaching traffic of your car's presence by signalling from well to the back of the hard shoulder, NOT the carriageway. DO NOT put yourself at risk. Face the approaching traffic, and be prepared to move quickly clear of the hard shoulder onto the verge if any vehicles pull onto the hard shoulder.
- **Get the passengers out of the car** – as a precaution should the car be hit by another vehicle. The passengers should move well away from the car and approaching traffic, up (or down) an embankment, or into a nearby field for example. If animals are being carried, it may be advisable to leave them in the vehicle. In any case, animals and children **must** be kept under tight control.
- **Call for assistance** – using the nearest emergency telephone on your side of the carriageway. NEVER cross the carriageways to use the emergency telephones.

  The direction of the nearest telephone is indicated by an arrow on the marker post behind the hard shoulder (the marker posts are positioned 100 metres/300 feet apart).

  Each telephone is coded with details of its location, and the operator will automatically be informed of your position on the motorway when you make the call.

  Give the operator details of your car type and registration number and, if you're a member of one of the motoring organisations, ask the operator to arrange assistance.
- **Don't leave your car unattended for a long period.**

Changing a wheel

- **First of all, make sure that the car is in a safe position.** It may be safer to drive the car slowly forwards and risk damaging the wheel, than to risk causing an accident.
- **If you're in any doubt** as to whether you can safely change the wheel without putting yourself or other people at risk, call for assistance.
- **All the tools required to change a wheel are kept in the boot.**
  However, if you're carrying any luggage, you'll need to move it for access to the spare wheel.

  Lift up the floor mat, and where applicable lift up the cover panel, to expose the spare wheel. On some models, the wheel is covered by a circular board cover, in which case a plastic retainer must be turned through 45° to remove the cover.

  The spare wheel is secured in place with a bolt, which must be unscrewed before the wheel can be lifted out.

On Saloon (Sapphire) and Hatchback models, the jack and the wheel brace (a spanner for unscrewing the wheel bolts) are attached to the rear panel in the luggage compartment, and are secured either by wing-nuts, or with rubber retainers. On some models, the jack and wheel brace may be covered by a panel which can be removed by turning a plastic fastener.

On Estate models, the jack and wheel brace are located under the floor cover panel, on the right-hand side next to the spare wheel. Turn the jack handle clockwise to free the wheel brace.

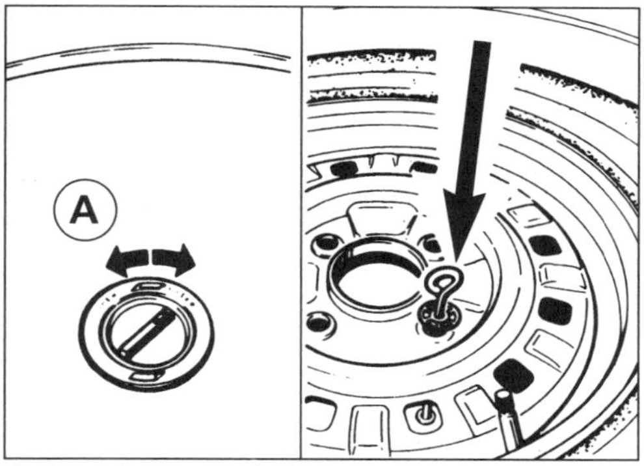

▲ *Spare wheel cover plastic retainer* ***(A)*** *and spare wheel securing bolt (arrowed)*

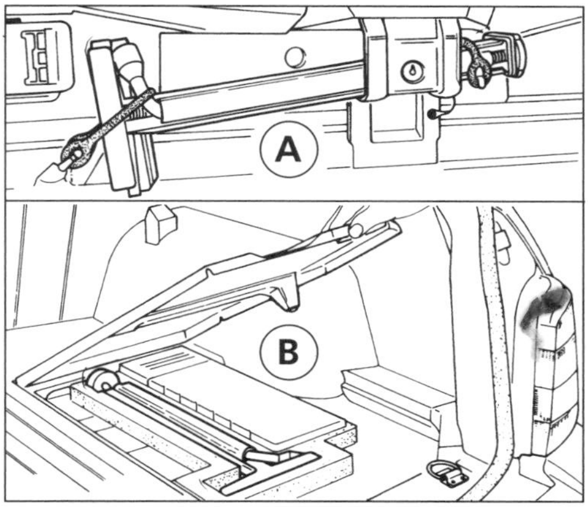

▲ *Jack and wheel brace location on Saloon and Hatchback models* ***(A)****, and Estate models* ***(B)***

- **Lift the spare wheel and the tools out of the boot,** and place them on the ground. Don't put any luggage back in the boot yet, because the wheel with the punctured tyre will have to be fitted in the spare wheel well.
- **Make sure that the handbrake is applied,** and select reverse gear on models with a manual gearbox, or position **P** on models with an automatic transmission.
- **Prise off the wheel trim (where applicable),** which covers the wheel bolts on the wheel with the puncture, using the blade-shaped end of the wheel brace.
- **Slacken the four wheel bolts** by at least half a turn, with the wheel on the ground, using the wheel brace. If you carry a length of piping for just such emergencies, slide it over the end of the wheel brace to make it easier to unscrew the bolts.
- **Before jacking the car,** make sure that the ground is solid enough to safely take the jack, and try to find something to chock the wheel diagonally opposite the one to be changed (two large stones or pieces of wood will do).
- **The jacking points** are shown in the accompanying illustration, and small cut-outs in the flange under the car show their locations.
- **Choose the jacking point nearest the wheel to be changed,** and engage the arm of the jack as far as it will go into the jacking point socket under the car. When the jack is correctly engaged, the hook on the end of the arm will prevent it from slipping out of the socket.

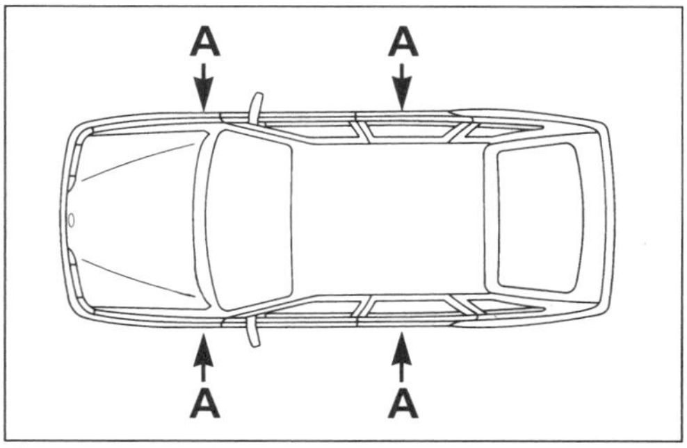

▲ *Jacking points* ***(A)*** *for use when wheel changing*

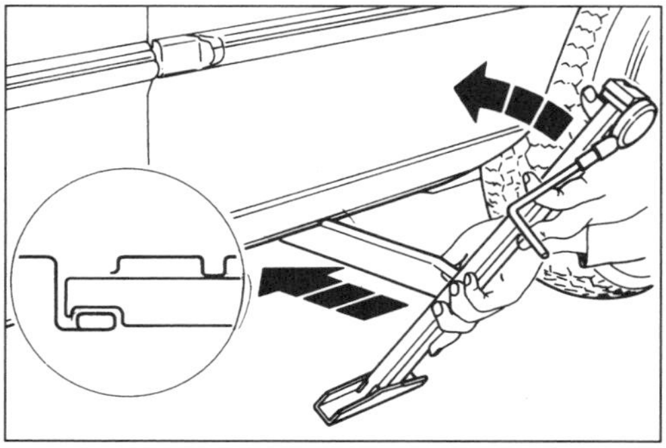

▲ *Using the car jack*

Unfold the jack handle, then turn the handle until the large plate on the bottom of the jack is resting evenly on the ground.

Carry on turning the jack handle slowly to raise the car until the wheel is clear of the ground, making sure that the car doesn't move on the jack.

Slide the spare wheel under the door sill panel for the time being, as near to the punctured wheel as possible. This will break the car's fall (and minimise the risk of injury to you) if the car slips off the jack or the jack collapses.

- **Unscrew all the wheel bolts,** then remove the wheel. Sometimes it may be necessary to tap the wheel in order to remove it, but if this is done, make sure that the car doesn't move on the jack.

  Remove the spare wheel from under the side of the car, and put the punctured wheel in its place for the time being.
- **Fit the spare wheel to the car,** and lightly tighten the wheel bolts in a crosswise order. Remove the punctured wheel from under the side of the car, then slowly lower the car to the ground.
- **Remove the jack, and fully tighten the wheel bolts.** There is no need to strain yourself doing this – just make sure that the bolts are reasonably tight.
- **Refit the wheel trim,** where applicable.
- **Stow the wheel** which has been removed in the boot, together with the jack and wheel brace.
- **Remove the wheel chocks,** and make a final check to ensure that all tools and debris have been cleared from the roadside.
- **Select neutral,** where applicable, before starting up and driving away.

Towing

If your car needs to be towed, or you need to tow another car, special towing eyes are provided at the front and rear of the car.

When being towed, the ignition key should be turned to position **II** (steering lock released, and ignition warning light on). This is necessary for the direction indicators, horn and brake lights to work.

Note that when being towed, the brake servo does not work (because the engine is not running). This means that the brake pedal will have to be pressed harder than usual to operate the brakes, so allowance should be made for greater stopping distances. If the breakdown doesn't prevent the engine from running, the engine could be started and allowed to idle so that the servo operates while towing.

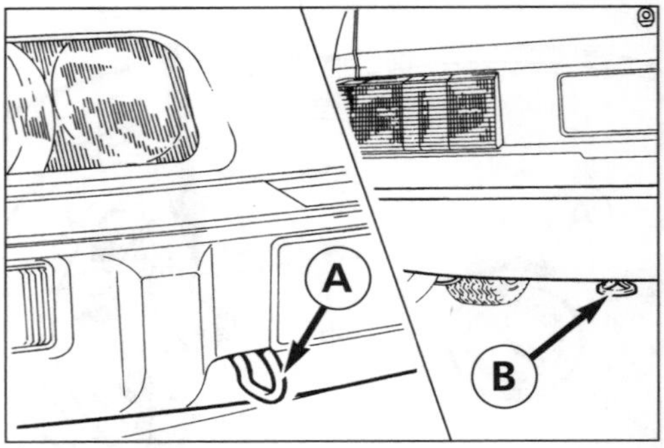

▲ *Front* ***(A)*** *and rear* ***(B)*** *towing eye locations*

When a car with an automatic transmission is being towed, the gear selector lever should be moved to the **N** position. To avoid damage to the transmission, do not tow the car any faster than 30 mph (50 km/h), or any further than 30 miles (50 km).

An 'On-tow' notice should be displayed prominently at the rear of the car on tow to warn other drivers.

Make sure that both drivers know details of the route to be taken before moving off, as it is difficult, and dangerous, to try to communicate once underway.

Before moving away, the tow car should be driven slowly forwards to take up any slack in the tow rope.

The driver of the car on tow should make every effort to keep the tow rope tight at all times, by gently using the brakes if necessary, particularly when driving downhill.

Drive smoothly at all times, particularly when moving away from a standstill. Allow plenty of time to slow down and stop when approaching junctions and traffic queues.

Starting a car with a flat battery

Apart from old age (most batteries should last for at least three years), a battery will normally only go flat if there is a fault in the charging circuit (indicated by a continuously glowing ignition warning light), or when a particular circuit (eg headlights) is left on for a long time with the engine switched off.

To start a car with a flat battery, you can either use a push or tow start (models with manual gearbox only), or a set of jump leads.

TOW/PUSH STARTING

The method used to start the engine is the same whether the car is being towed or being pushed but, if the car is to be towed, first read the previous Section on towing for details of where to connect the tow rope.

Proceed as follows:

1. Turn the ignition key to position **II** to switch on the ignition.
2. If the engine is cold, on models with a manual choke, pull out the choke control.
3. Depress the accelerator pedal.
4. Depress the clutch pedal and select third gear. Hold the clutch pedal down.
5. Tow or push the car, and slowly release the clutch pedal. The engine will turn over, and should start (don't worry about the car 'juddering' as the engine turns). If the car is being towed, as soon as the engine starts, depress the clutch pedal and *gently* brake the car so that you don't run into the tow vehicle.

STARTING USING JUMP LEADS

Note: *Starting using jump leads can be dangerous if done incorrectly. If you're uncertain about the following procedure, it's recommended that you ask someone suitably qualified to do the jump starting for you.*

Before attempting to use jump leads, there are a few important points to note.

First of all, use only proper jump leads which have been specifically designed for the job.

Make sure that the booster battery (the fully-charged battery) is a 12-volt type.

Position the two vehicles close enough together to connect the leads, but **do not** allow the vehicles to touch.

Turn off all electrical circuits.

DO NOT allow the ends of the two jump leads to touch at any time during the following procedure.

Proceed as follows.

1. Open the vehicle bonnets, and connect one end of one of the jump leads (usually the red one) to the positive (+) terminal of the flat battery, and the other end to the positive terminal of the booster battery.
2. Connect one end of the other jump lead to the negative (-) terminal of the booster battery, and connect the other end to a suitable earth point on the car with the flat battery, at least 50 cm (18 in) away from the battery (eg a clean, bare metal area on the engine block or body). Make sure that all the lead clips are secure.
3. Start the engine of the vehicle with the booster battery and let it run for a few minutes, then start the engine of the car with the flat battery in the normal way.
4. When the engine is running smoothly, disconnect the jump leads in exactly the reverse order to that in which they were connected.

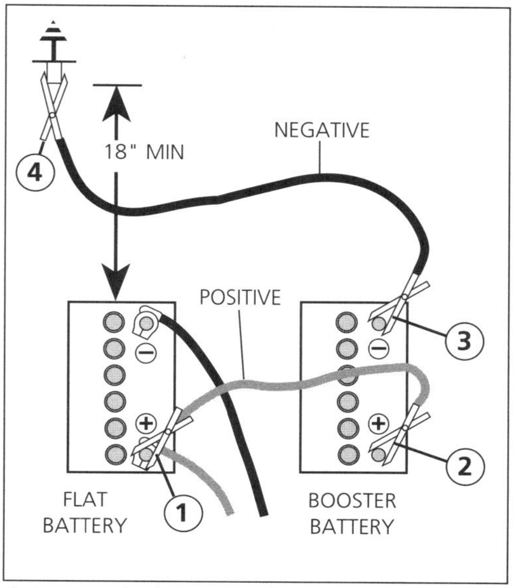

▲ *Jump start lead connections for negative earth vehicles – connect leads in order shown*

What to carry in case of a breakdown

Carrying the following items may help to reduce the inconvenience and annoyance caused if you're unlucky enough to break down during a journey:

- Alternator drivebelt (refer to *'Servicing'* on page 102 for details of renewal)
- Spare fuses (refer to *'Bulb, fuse and relay renewal'* on page 119 for details of renewal)
- Set of principal light bulbs (refer to *'Bulb, fuse and relay renewal'* on page 119 for details of renewal)
- Battery jump leads (refer to *'Starting a car with a flat battery'* on page 47 for details of starting using jump leads)
- Tow-rope (refer to *'Towing'* on page 46 for details of towing procedure)
- Litre of engine oil (refer to *'Regular checks'* on page 74 for details of checking oil level)
- Torch

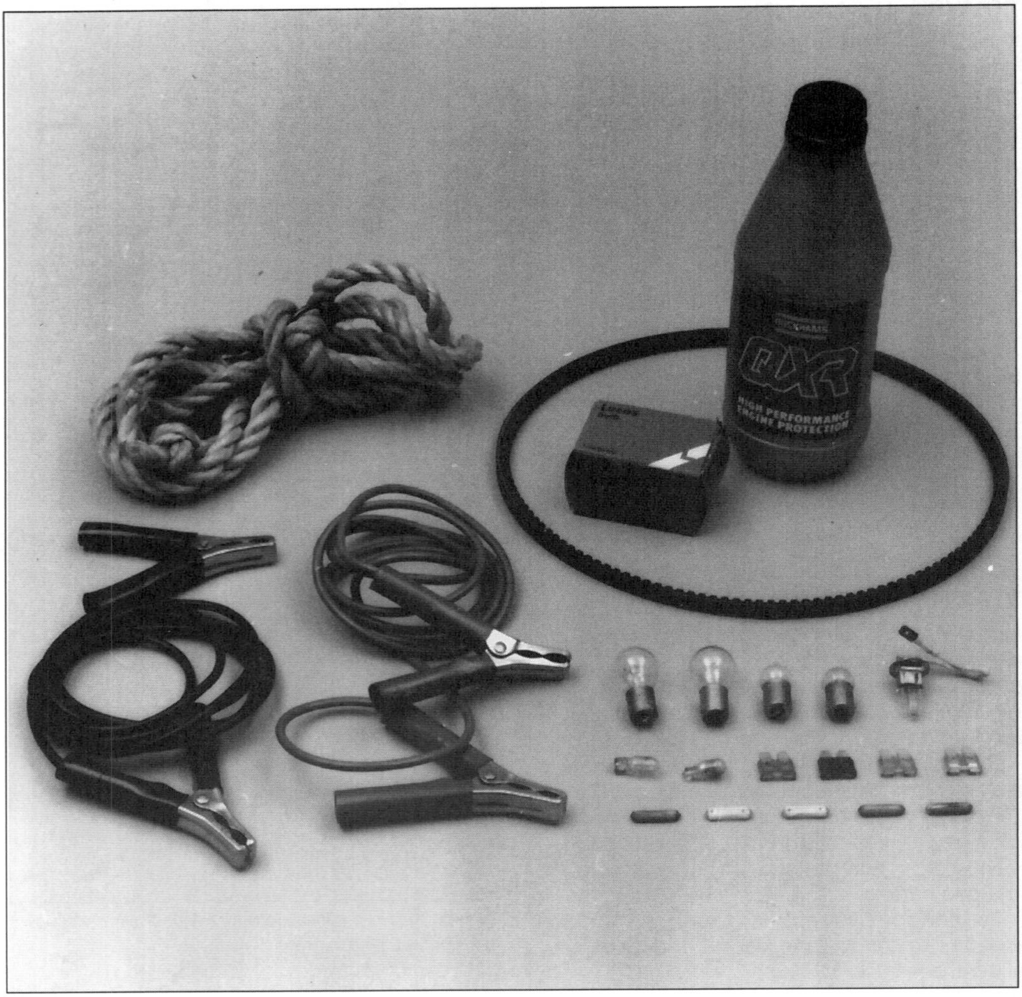

In the UK, more people are injured or killed every year due to road accidents than through any other single cause.

Car manufacturers are paying more attention to passive safety when designing new cars, but drivers must ultimately take active responsibility of ensuring the safety of themselves, their passengers, and just as importantly, other road users.

There will always be accidents on the roads, but taking the time to read the following advice may help you to avoid such a mishap.

The following Sections aim to give advice which will help you to reduce the risk of becoming involved in an incident when driving, not necessarily by changing the way that you drive, but simply by making you aware of some of the potential risks which can easily be avoided. For more information and practical advice on road safety, it is well worth considering taking part in one of the various courses provided by organisations such as RoSPA and the Institute of Advanced Motorists.

BEFORE STARTING A JOURNEY

As a driver, before beginning a journey it's important to make sure that you're comfortable so that you can concentrate on driving without any unnecessary distractions. Spending a few moments carrying out a few simple checks and adjustments will help to make your journey more relaxing, safer and hopefully trouble-free.

Driver comfort

Before driving, make sure that you're comfortable, and that you can operate all the controls easily, particularly if someone else has recently driven the car.

- **Seat**
  Make sure that the seat is positioned a comfortable distance from the steering wheel and the pedals, so that you can operate all the controls comfortably
- **Seat back**
  Make sure that the seat back is adjusted to give your back plenty of support. Your back should rest against the seat, and there should be no need to lean forwards from the seat back. You should be able to reach the steering wheel comfortably, and you should be able to turn the wheel easily without stretching.
- **Head restraint**
  If a head restraint is fitted, it should be adjusted to support your head in the event of an accident. Your head should not rest against the restraint during normal driving and, as a rough guide, the restraint is correctly positioned when its top is in line with your eyes. Similarly, make sure that any passenger head restraints are adjusted correctly if passengers are being carried.
- **Mirrors**
  All the mirrors should be adjusted so that they give a clear view behind the car without the need to move your head unnecessarily.
- **Steering column**
  If the car is fitted with an adjustable steering column, adjust the column so that the steering wheel can be reached comfortably, and the wheel can be turned easily without stretching.

Checks

If you're planning a long journey, refer to *'Regular checks'* on page 70, and make sure that you carry out all the checks mentioned before setting off. Also make sure that you know where the spare wheel and the tools required for wheel changing are located (refer to *'Breakdowns'* on page 43).

DRIVING IN BAD WEATHER

When driving in bad weather, always be prepared. Bad weather should never take you by surprise. Listen to a weather forecast: you should always be aware of the possibility of poor conditions on your intended journey.

Driving in bad weather requires more concentration, and is more tiring than driving in good weather conditions. Never allow yourself to be distracted by talkative passengers or loud music in the car and, if you feel tired, stop at the next opportunity and take a break.

Always look well ahead, so that you're aware of the condition of the road surface, and any obstacles; **slow down if necessary**.

The following advice is intended to help you to drive more safely in various bad weather conditions. However, above all, use common sense.

Rain

- Use dipped headlights in poor visibility.
- Slow down if visibility is poor or if there is a lot of water on the road surface.
- Keep a safe distance from the vehicle in front (stopping distances are doubled on a wet road surface).
- Be particularly careful after a long period of dry weather. Under these conditions, rain can make the road surface very slippery.
- Don't use rear foglights unless visibility is **seriously** reduced (generally less than 100 metres/300 feet). Foglights can dazzle drivers following behind, especially in motorway spray.

Fog

- Slow down. Fog is deceptive, and you may be driving faster than you think. Fog can also be patchy, and the visibility may suddenly be reduced. Always drive at a speed which allows you to stop in the distance you can see ahead.
- Use dipped headlights. Using main-beam headlights will usually reduce the visibility even further, as the fog will scatter the light.
- Use foglights (where they are fitted) if it's genuinely foggy (generally, where visibility is reduced to less than 100 metres/300 feet), and not just misty.
- Keep a safe distance from the vehicle in front.
- Use your windscreen wipers to clear moisture from the windscreen.

Snow

- Don't start a journey if there is any possibility that conditions may prevent you from reaching your destination. Listen to a weather forecast before setting off.
- Before starting a journey, clear **all** snow from the windscreen, windows and mirrors. Don't just clear a small area big enough to see through.
- Slow down.
- Keep a safe distance from the vehicle in front (stopping distances can be trebled or even quadrupled on a snow-covered or icy surface).
- Drive smoothly and gently. Accelerate gently, steer gently and brake gently.
- Don't brake and steer at the same time – this may cause a skid (refer to *'Skid control'* on page 54).
- When moving away from a standstill, or manoeuvring at junctions, etc, in a car with a manual gearbox use the highest possible gear that your car will accept to move away, and change up to a higher gear earlier than

usual. In a car with automatic transmission move the selector lever to prevent the transmission from using all the gears (usually position '2' – refer to *'Controls and equipment'* on page 14). This will provide more grip and will help to reduce wheelspin.

- Use main roads and motorways where possible. Major roads are likely to have been 'gritted' and are usually cleared before minor roads.
- Use dipped headlights in poor visibility.
- Don't use rear foglights unless visibility is **seriously** reduced (generally less than 100 metres/300 feet) – foglights can dazzle drivers following behind.

Ice and frost

- Follow the advice given for driving in snow.
- Be prepared for 'black ice'. Although you can't see 'black ice', reduced road noise from the tyres will usually tell you it's there.

Severe winter weather

The best advice in severe winter weather is to stay at home but, if you must drive your car, in addition to following the advice given for driving on snow and ice, there are a few additional pieces of advice which you should bear in mind before setting out on a journey.

- Make sure that you tell someone where you're going, and tell them roughly what time you're expecting to arrive at your destination, and what route you're taking.
- Keep a can of de-icer fluid, a scraper, a set of jump leads and a tow rope in the car at all times.
- Carry plenty of warm clothes and blankets.
- Make sure that you have a full tank of petrol before starting your journey. This will allow you to keep the engine running, without fear of running out of petrol, to provide warmth through the car's heating system should you be delayed by conditions, or worse still stuck.
- Pack some pieces of old sacking or similar material, which you can place under the driving wheels to give better traction if you get stuck.
- Carry a shovel, in case you need to dig yourself or someone else out of trouble.

MOTORWAY DRIVING

Motorway driving requires a great deal of concentration and awareness. Motorways are generally busier than ordinary roads, and the traffic moves faster, so there is less time to react to any changes in road conditions ahead.

This Section covers the fundamental rules for safe motorway driving, and will help you to avoid many of the problems which occur on today's busy motorways. Remember that lack of common sense is probably the biggest cause of accidents on motorways.

Full rules and regulations for motorway driving can be found in 'The Highway Code'.

Joining and leaving a motorway

When you join a motorway, you will normally approach from a 'slip-road' on the left. You must give way to traffic already on the motorway. Watch for a safe gap in the traffic, and adjust your speed in the acceleration lane so that when you join the left-hand lane of the motorway you're already travelling at the same speed as the other traffic. Indicate before pulling onto the motorway. After joining the motorway, allow yourself some time to get used to the speed of the traffic before overtaking.

When you leave a motorway, indicate in plenty of time, and reduce your speed before you enter the slip-road – some slip-roads and link-roads between motorways have sharp bends which can only be taken safely by slowing down. Be very cautious immediately after leaving a motorway, as it can be difficult to judge speed after a long period of fast driving.

Rules for safe motorway driving

- **Concentrate and think ahead** – Always be prepared for the unexpected, and look well ahead. You should be aware of all the traffic in front and behind, not just the vehicle in front of you. If you're concentrating properly, nothing should take you by surprise on a motorway.
- **Always drive at a speed to suit the road conditions** – Don't break the speed limit, especially temporary speed limits which may apply to contra-flow systems or roadworks.
- **Slow down** in bad weather conditions. Driving too fast in fog and motorway spray is a major cause of motorway accidents.
- **Always keep a safe distance from the vehicle in front** – The closer you drive to the vehicle in front, the less chance you have of avoiding an accident if something happens ahead. Remember that you need more space in bad weather conditions.
- **Think** – If the vehicle ahead suddenly stops, do you have enough space to stop without hitting it?
- **Use your mirrors regularly** – It's just as important to be aware of what's happening behind you as it is in front.
- **Always signal your intentions clearly and in plenty of time** – Check your mirrors first, signal *before* you move, and change lanes smoothly and in plenty of time. Don't make any sudden moves which are likely to affect traffic approaching from behind.
- **Keep to the left** – The left-hand lane is *not* the slow lane, and you should drive in this lane whenever possible. You can stay in the middle lane when there are slower vehicles in the left-hand lane, but return to the left-hand lane when you've passed them. The right-hand lane is for overtaking only: it is *not* for driving at a constant high speed. If you use the right-hand lane, move back into the middle lane and then into the left-hand lane as soon as possible, but without cutting in.
- **Take note of direction signs** – You may need to change lanes to follow a certain route, in which case you should do so in plenty of time.
- **Stop as soon as practicable if you feel tired** – Driving on a motorway when feeling tired can be extremely dangerous. If necessary, wind down the window for fresh air. If you feel tired or drowsy, stop at the next service station, or turn off the motorway at the next junction and take a break before continuing your journey. You must not stop on the hard shoulder of a motorway other than in an emergency.
- **If you break down** – Refer to *'Breakdowns'* on page 43.

TOWING A TRAILER OR CARAVAN

When towing a trailer or a caravan, there are a few special points to bear in mind. There's more to towing a trailer or caravan than just simply bolting on a towbar and hitching up!

The law

Before using your car for towing, make sure that you're familiar with any special legislation which may apply, particularly if you're travelling abroad.

Make sure that you know the speed limits applicable for towing.

In some countries, including the United Kingdom, there is a legal requirement to have a separate warning light fitted in the car to show that the trailer/caravan direction indicator lights are working.

Always check on current regulations before you travel.

Before starting a journey

Before attempting to use your car for towing, there are one or two points which should be considered.

- **Make sure that your car can cope with the load which you're towing**
- **Don't exceed the maximum trailer or towbar weights for the car** – Refer to *'Dimensions and weights'* on page 12.
- **Engine** – Don't put unnecessary strain on your car's engine by trying to tow a very heavy load. Obviously, the smaller the engine, the less load the car can comfortably tow. Also bear in mind that the extra load on the engine when towing means that the engine's cooling system may no longer be adequate. Some manufacturers provide modified cooling system components for towing, such as larger radiators, etc, and if you intend to use your car for towing regularly, it may be worth enquiring about the availability and fitting of these components.
- **Suspension** – Towing puts extra strain on the suspension components, and also affects the handling of the car, since standard suspension components aren't usually designed to cope with towing. Heavy duty rear suspension components are available for most cars, and it's worth considering having these fitted if you intend to use your car for towing regularly. Special stabilisers are also available for fitting to most cars to reduce pitching and 'snaking' movements when towing.
- **Make sure that you can see behind the trailer/caravan using the car's mirrors** – Additional side mirrors with extended arms are available to fit most cars. The mirrors must be fitted with folding arms (so that they fold if hit), and should be adjusted to give a good view to the rear at all times. The mirrors can be removed when the car isn't being used for towing.
- **Make sure that the tyre pressures are correct** – Unless a light, unladen trailer is being towed, the car tyres should be inflated to their 'full load' pressures. Also make sure that where applicable the trailer or caravan tyres are inflated to their recommended pressure.
- **Make sure that the headlights are set correctly** – Check the headlight aim with the trailer/caravan attached, and if necessary adjust the settings to avoid dazzling other drivers.
- **Make sure that the trailer/caravan lights work correctly** – Make sure that the tail lights, brake lights, direction indicator lights and, where applicable, the reversing lights and fog lights all work correctly. Note that the additional load on the direction indicator circuit may cause the lights to flash at a rate slower than the legal limit, and in this case you may need to fit a 'heavy duty' flasher unit.
- **Make sure that the trailer/caravan is correctly loaded** – Refer to the manufacturer's recommendations for details of loading. As a general rule, distribute the weight so that the heaviest items are as near as possible to the trailer/caravan axle. Secure all heavy items so that they can't move.
  To provide the best possible control and handling of the car when towing, the manufacturers recommend an optimum noseweight for the trailer/caravan when loaded (refer to *'Dimensions and weights'* on page 12 for details). The noseweight can be measured using a set of bathroom scales as follows:
  Place a stout piece of wood between the trailer/caravan tow hitch cup and the scales platform, and read off the weight with the trailer/caravan level. If necessary, redistribute the load, to arrive as close as possible to the recommended noseweight. **Do not** exceed the maximum recommended noseweight.

Driving tips

Towing a trailer or caravan will obviously affect the handling of the car, and the following advice should be followed when towing.

- **If possible, avoid driving with an unladen car and a loaded trailer/caravan** – The uneven weight distribution will tend to make the car unstable. If this is unavoidable, drive slowly to allow for the instability.

- **Always drive at a safe speed** – The stability of the car and the trailer or caravan decreases as the speed increases. Always drive at a speed which suits the road and weather conditions, and always reduce speed in bad weather and high winds – especially when driving downhill. If the trailer or caravan shows any sign of 'snaking', reduce speed immediately – never try to stop 'snaking' by accelerating.
- **Always brake in good time** – If you're towing a trailer or caravan which has brakes, apply the brakes gently at first, then brake firmly. This will help to prevent the trailer wheels from locking. In cars with a manual gearbox, change into a lower gear before going down a steep hill so that the engine can act as a brake, and similarly, in cars with automatic transmission, move the selector lever to position '2', or '1' in the case of very steep hills.
- **Don't change to a lower gear unnecessarily** – Unless the engine is labouring, stay in as high a gear as possible to keep the engine revs as low as possible. This will help to avoid the engine overheating.

ALCOHOL AND DRIVING

By far the best advice on drinking and driving is **DON'T**.

You may feel fine, but it's a proven fact that even one small alcoholic drink will impair your driving to some extent.

Drinking alcohol has the following effects:

- **Reduces co-ordination**
- **Increases reaction time**
- **Impairs judgement of speed, distance and risk**
- **Encourages a false sense of confidence**

The risk of an accident increases sharply after drinking alcohol, and approximately one third of the total number of people killed in road accidents each year have blood alcohol levels above the legal limit for driving. Remember that you're putting other people as well as yourself at risk if you drive after drinking.

The legal limit for blood alcohol level in the UK is 80 milligrams per 100 millilitres. This doesn't correspond to any particular quantity of drink, as the amount of drink required to reach this level varies from person to person. The driving of many people who feel perfectly sober is seriously affected well below the legal limit.

The penalties for driving over the legal limit are severe, and can mean losing your driving licence, a heavy fine, or imprisonment. It's also important to realise that the laws abroad can be far more severe than in the UK, and some countries have a total ban on driving after drinking alcohol.

- **The safest course of action is not to drink and drive.**

SKID CONTROL

If you drive sensibly with due regard for the road conditions, you should never find yourself in a situation where your car is skidding. The following advice will help you to avoid situations which may cause a car to skid, and explains how to regain control of your car quickly should the need arise.

A skid is caused by one or a combination of the following:

- **Excessive speed in relation to the road conditions.**
- **Harsh or excessive acceleration.**
- **Sudden or excessive braking.**
- **Coarse or excessive steering.**

The most common basic cause of skidding is rough handling of the car's controls. Therefore the key to safe driving and preventing a skid is smooth, gentle handling of the car and its controls. Try to apply smooth pressure to the brake pedal, accelerator and steering wheel rather than just suddenly moving them a certain distance.

When a car is skidding, the following things happen:

- **The car is out of control. You can take action to regain control, but while skidding, the car can't be fully controlled.**
- **A car with the front wheels skidding can't be steered.**
- **A car with the rear wheels skidding is likely to spin round if any steering lock is applied.**
- **A car with all four wheels skidding will continue in a straight line in the direction it was travelling when the skid started, regardless of which way the car is pointing.**

Prevention is better than cure. When approaching a corner, reduce speed early by smooth progressive braking **in a straight line**. In a car with a manual gearbox, brake to an appropriate speed and select the correct gear for the corner. Turn into the corner early and smoothly, gently increasing the steering lock if necessary, and corner with your foot gently on the accelerator, so that the engine just 'pulls' the car through the corner, maintaining a constant speed. As you leave the corner, accelerate gently and smoothly.

If you follow the above advice, you should avoid finding yourself in a situation where your car is skidding. However, should you be unfortunate enough to experience skidding, the basic rule for skid control is to **remove the cause** of the skid. If you cause a skid by accelerating, braking or steering, **ease off**, and, when the skid stops, re-apply the accelerator, brake or steering control, but this time more gently and smoothly. If the front wheels are skidding, to regain steering control quickly, steer 'into the skid'. This is often misunderstood, and it basically means look in the direction you want to be facing, and steer in that direction – and be prepared to reduce the steering quickly if necessary when the wheels regain their grip.

Several organisations offer courses in skid control, using specially-equipped cars under carefully-controlled conditions. One of these courses could prove to be a very worthwhile investment, possibly helping you to avoid an accident.

ADVICE TO WOMEN DRIVERS

Unfortunately, it's a fact that women are more likely than men to be the subject of unwelcome attention when driving alone.

There's no reason to think that driving alone spells trouble, but it's a good idea to be aware of certain precautions which will help to avoid finding yourself in an unpleasant situation.

This Section aims to give advice which will help to reduce any risks when driving alone, and will help you to deal with the situation, should you be unfortunate enough to find yourself the subject of unwelcome attention.

Several organisations (some local police authorities, AA, etc) run courses especially for women drivers, in which subjects such as basic car maintenance and self defence, etc, are covered.

Sensible precautions

● The car

Make sure that you're familiar with your car.

Make sure that your car is regularly serviced. Refer to *'Regular checks'* on page 70, and make sure that all the checks described are carried out regularly. This will minimise the possibility of a breakdown.

Make sure that you know how to change a wheel, and make sure that the car jack is in good condition in case you have a puncture – refer to *'Breakdowns'* on page 43.

● The journey

If you're planning to undertake a particularly long or unfamiliar journey, the following advice may be helpful.

If you're travelling on an unfamiliar route, make a few notes before you set off, reminding yourself of the road numbers, where to turn, which junctions to use on motorways, etc. Try to stick to main roads where possible.Always carry a map.

Make sure that you have enough petrol for the journey. If you need to fill up during the journey, make sure that you do so in plenty of time, before the fuel level gets too low, and before petrol stations close if travelling at night.

If you think that your family or friends may be worried, you may want to 'phone someone at your destination before setting off to tell them that you're leaving, and tell them what time you expect to arrive. Similarly, when you arrive, you may want to 'phone someone at your starting point to confirm that you've arrived safely.

● Driving in traffic

Avoid attracting unnecessary attention.

Avoid 'jokey' stickers in car windows, which may encourage unwelcome attention.

Avoid eye contact with 'undesirables'.

If you're being followed, pull over and slow down, but **don't** stop. **Don't** drive to your home, but find a busy and well lit place. If the person following persists, blow your horn and flash your lights to attract attention.

If you're stopped by traffic or another vehicle, lock the doors and close the windows. **Don't** ram the other vehicle – damage to your car might prevent your escape.

● Leaving your car

Don't leave any clues that the car is being driven by a woman; make sure that any 'feminine' items are hidden from sight before leaving your car.

Refer to *'Car crime prevention'* on page 65 and take note of the advice given.

Try to park in a well-lit, preferably busy area.

If you park in a car park, try to park close to an exit, or close to the attendant's station. **Always** reverse into a parking space, so that you can drive away quickly if necessary after returning to your car.

Take note of any 'landmarks' so that you can find your car quickly when you return.

Always lock your car.

When you return to your car, walk with a group of people, if possible. Have the keys ready so that you don't have to spend unnecessary time outside your car searching for them, which may attract attention.

Before getting into your car, briefly check for forced entry, and look into the car for any suspicious signs. **Don't** get into the car if you notice anything suspicious.

● Breakdowns

Remember that you're more likely to be injured in an accident than a personal attack.

Refer to *'Breakdowns'* on page 43 and take note of the advice given but, in addition, bear in mind the following advice.

Always walk to 'phone for assistance, **don't** accept a lift. When you 'phone for assistance, mention that you're an unaccompanied woman.

If someone stops while you're 'phoning for help, give the operator details of the other vehicle's registration number and a brief description of the car and driver. If the driver approaches you, tell them that you have passed on his/her details to the police. If the driver's intentions are honourable, your reaction will be understood.

If you decide to stay with, but outside the car, leave the nearside door unlocked and slightly open, so that you can get inside quickly to lock yourself in if necessary.

If you decide to stay inside the car, sit in the passenger seat and lock the door – this will give the appearance that you're accompanied.

When help arrives, ask for some form of identification (even from a policeman) before giving any of your own details.

If someone offers assistance, tell them the police have been informed and are arranging recovery. If you have not yet contacted the police, consider asking the person to do so on your behalf, **but** if you're uncertain about the person's intentions, tell them that the police are aware and ask them to call the police again for you.

● **Accidents**

Keep calm.

Refer to *'Accidents and emergencies'* on page 35 and take note of the advice given but, in addition, bear in mind the following advice.

If you're bullied or shouted at, lock yourself in your car (if it's safe to do so), and wind the windows up. Communicate through a small gap at the top of the window.

If things get out of hand, refuse to talk to anyone except the police.

Self defence

This is a last resort and should not be used unless all else has failed.

There is no substitute for taking a properly supervised course in self-defence, but we aim to outline a few of the basic moves below.

If you are approached, act in a composed manner and use a soft tone.

In the event of an attack, stay calm – you can still safeguard yourself.

- **If you're held by one arm** – Use your free hand to grasp the attacker's thumb and twist the thumb sharply back towards the wrist.
- **If you're facing your attacker** – If your shoulders are held, thrust your hands/arms upwards and diagonally outwards to dislodge the attacker's grip.

  Use your knees or the blade of your hand to chop your attacker's groin, your feet to kick the shins, and your fingers to poke the eyes.
- **If you're attacked from behind** – Quickly lean forward, twisting your head sideways into the attacker to keep your airway (for breathing) clear. Often this will cause the attacker to lose balance.
- **If you've been pulled backwards already** – Turn your head as above, and chop hard with the blade of your hand or a clenched fist to the groin, or an elbow in the stomach.
- **If you've been forced to the floor** – Use your feet and legs to kick against the attacker's shins whilst swivelling your body to keep the attacker at bay.

Be ready to run as soon as the attacker's grip is released, and SCREAM!

CHILD SAFETY

Every day, thousands of parents strap their offspring into child car seats, confident that they have done everything possible to protect their loved ones. However, a recent survey has shown that many young children are travelling in car seats which have been incorrectly fitted or are being incorrectly used, making them potentially dangerous should they be involved in an accident. The following advice will help you to make sure that you take every possible precaution to ensure the safety of your children when driving.

How to avoid unnecessary risks

- **Never allow young children to travel in a car unrestrained,** even for the shortest of journeys.
- **Never carry a child on an adult's lap or in an adult's arms.** Although you may feel that your baby is safer in your arms, this is not the case. An adult holding a child is far more likely to cause injury to the child than to give protection in the event of an accident.

- **Never rely solely on an adult seat belt to restrain a child,** and never sit a child on a cushion to enable a seat belt to fit properly.
- **Always strap young children into a properly designed child car seat** when carrying them in a car. The cost of a good car seat is a very small price to pay to save your child's life.

Choosing a child car seat

- It's vitally important to choose the correct type of car seat for your child. A wide range of child car seats is available, and it's worth spending some time looking at the various seats on the market before deciding on the most suitable seat for your particular requirements.
- Although age ranges are often given by the manufacturers, these should be taken as a rough guide. It's the weight of the child which is important; for example, a smaller than average baby could use a baby car seat for longer than a heavy baby of the same age.
- **Never** buy a secondhand child seat. This may sound like a ploy from the manufacturers to sell more seats, but all too often a secondhand seat is sold without the instructions, and sometimes there are parts missing. This often leads to secondhand seats being incorrectly fitted. A secondhand seat may have been damaged or weakened through carelessness or misuse, without necessarily showing any visible signs until it's put to the test in an accident!
- Your baby's first contact with the outside world is often on that first ride home from hospital, so make sure that you're prepared, and buy a car seat before your baby is born.
- Some seats are designed for babies from birth to a weight of around 22 pounds (10 kg), which for most babies is around nine months old. Usually, these first car seats are light and easy to transport through the use of a handle. This means that a sleeping baby can be carried from the car into the house without waking. Some baby seats come complete with a built-in headrest, which is very important for the early weeks when a baby's head needs to be supported.

  Certain seats can be fitted so that the baby faces the back of the car (ie rearward facing seats). This may be considered to improve safety, as the baby is supported across the back rather than purely by the harness, if the car is involved in a frontal impact. Combination type seats can then be used forward facing when the child reaches approximately nine months old.

  Other types of seat may be designed to accommodate children up to around 40 pounds (18 kg) or approximately four years old.

  Alternatively, some seats use the car's seat belts to hold both the child seat and the child, and these seats are obviously easier to fit. Make sure that this type of seat is fitted with a seat-belt lock, so that the seat belt cannot be pulled out of place or slackened. Another advantage of these seats is that they can easily be transferred from car to car.
- Try to choose a seat which has an easily adjustable harness, as this makes it easier to ensure that the harness fits the child securely for each trip. If the harness can be quickly loosened, it makes getting a struggling child in and out a little easier.

Using a child car seat

- Firstly, make absolutely sure that the seat is properly fitted in accordance with the manufacturer's instructions. If you're unsure about any of the fitting procedure, contact the manufacturer for advice.
- To hold a child securely, the harness must be reasonably tight. There should be just enough room to slide your flat hand under the strap. Children may be wearing bulky clothes one day, and thin clothes the next, so it is vital that the harness is adjusted before each journey to ensure a correct fit.
- If the seat is fitted using an adult's inertia reel seat belt, make sure that the seat is held firmly in position, and that the seat belt is securely locked to prevent it from loosening. Also, make sure that the seat belt buckle is not resting on the frame of the child seat. This is because the buckles are not designed to withstand the impact of a heavy child seat and, in an accident, could break open.

Driving abroad can be very different to driving in the UK. Besides the obvious differences, like climate and driving on the right-hand side of the road, various unfamiliar laws may apply, and it's advisable to prepare yourself and your car as far as possible before travelling.

This Section provides you with a guide which will help you to prepare for driving abroad, and will help you to avoid some of the pitfalls waiting for the unwary.

It may be worthwhile considering hiring a car abroad, rather than driving your own car. In this case, make sure that the insurance cover arranged suits your requirements, and make sure that a damage waiver is included (otherwise you will have to pay for any damage to the hire car).

Insurance

● Motoring

Make sure that you have adequate insurance cover for your car and your luggage.

Check on the legal requirements for insurance in the country you're visiting, and always inform your insurance company that you're taking your car abroad – they will be able to advise you of any special requirements.

Most car insurance policies automatically give the minimum legally-required cover for driving in EC countries, but if you require the same level of cover as you have in the UK, you will normally need to obtain a 'Green Card' (an internationally-recognised certificate of insurance) from your insurance company.

● Medical

It's always advisable to take out medical insurance for the car occupants. Not all countries have a free emergency medical service, and you could find yourself with a large unexpected bill in the event of yourself or one of your passengers being taken ill, or being involved in an accident (in some countries you may even have to pay for an ambulance).

NHS form E111 (available by filling in a form at a Post Office) entitles you to receive the same health care that residents receive in EC countries, but this is by no means comprehensive, and additional insurance cover is usually advisable.

● **Breakdown**

Recovery and breakdown costs can be far higher abroad than in the UK.

Most of the national motoring organisations (such as the AA and RAC) will be able to provide insurance cover which could save you a lot of inconvenience and expense should you be unfortunate enough to break down.

Documents

Always carry your passport, driving licence, car registration document, and insurance certificate (including 'Green Card' and medical insurance, where applicable).

Make sure that all the documents are valid, and that the car's road fund licence (tax) and MOT don't run out while you're abroad.

Before travelling, check with the authorities in the country you're visiting, in case any special documents or permits are required. You may need a visa to visit some countries, and an International Driving Permit (available from the AA or RAC) is sometimes required.

Driving laws

Before driving abroad, make sure that you're familiar with the driving laws in the country you're visiting, as there may be some laws which don't apply in the UK, and the penalties for breaking the law may be severe.

Fit a 'GB' plate to the back of your car, and make sure that it's displayed all the time you're abroad.

In some countries, you're legally required to carry certain items of safety equipment. These can include a first-aid kit, a warning triangle, a fire extinguisher, a set of spare bulbs/fuses, etc.

Remember that if you're visiting a country where you have to drive on the right-hand side of the road, you'll need to fit headlight beam deflectors or shields to avoid dazzling other drivers, or alternatively, it may be possible to have the headlight beams adjusted.

Make sure that you're familiar with the speed limits, and note that in some countries there is an absolute ban on driving after drinking *any* alcohol.

Servicing

Service your car before setting off on your trip, to reduce the possibility of any unexpected breakdowns.

Pay particular attention to the condition of the battery, windscreen wipers and tyres (including the spare), noting that the tyre pressures will probably have to be increased from their normal setting if the car is to be fully loaded. If you're travelling to a cold country, check the condition of the cooling system and the strength of the antifreeze.

Before setting off on your journey, refer to '*Regular checks*' on page 70, and carry out all the checks described. Check that the jack and wheel brace are in place in the car (refer to '*Breakdowns*' on page 43), and that the jack works properly.

Spares

In addition to the items which must be carried by law in the country you're visiting, it's a good idea to carry a few spares which may be difficult to obtain should you need them abroad (one of the national motoring organisations should be able to advise you). For example, you may want to carry clutch and throttle cable repair kits, as right-hand-drive components can be difficult to find outside the UK.

If your car uses a special oil, it's a good idea to take a pack with you.

It's also a good idea to carry a tow rope and a set of jump leads to help you out in case of a breakdown.

Fuel

The type and quality of petrol available varies from country to country, and it's a good idea to check on the availability of the correct petrol type before travelling (again, one of the national motoring organisations will be able to advise you). This is especially important if your car has a catalytic converter, as in this case you must only use unleaded petrol.

Find out what petrol pump markings to look for to give you the correct type and grade of petrol for your car.

Security

A foreign car packed with luggage is an inviting prospect for criminals, so refer to '*Car crime prevention*' on page 65, and don't take any chances.

Route planning

It's always a good idea to plan your approximate route before travelling. A vast number of maps and guides are available, or the AA and RAC can provide you with directions to your destination for a modest charge.

Bear in mind that in some countries, you'll have to pay tolls to use certain roads, and this can add unexpectedly to the cost of travelling.

What to carry when driving abroad

The following list provides a guide to the items which it is compulsory to use or carry, or it is strongly recommended that you carry in your car when driving in various European countries.

In the following table **'C'** indicates **'Compulsory'** and **'R'** indicates **'Recommended'**.

COUNTRY	HEADLAMP DEFLECTORS	GB STICKER	SEAT BELTS	WARNING TRIANGLE	FIRE EXTINGUISHER	FIRST AID KIT	SPARE BULBS
Austria	C	C	C	C	-	C	-
Belgium	C	C	C	C	C	-	-
Bulgaria	C	C	C	C	C	C	-
Czechoslovakia	C	C	C	C	-	C	C
Denmark	C	C	C	R	-	-	-
Eire	-	C	C	-	-	-	-
Finland	C	C	C	C	-	-	-
France	C	C	C	C	-	-	R
Germany	C	C	C	C	C	C	R
Greece	C	C	C	C	C	C	-
Holland	C	C	C	C	-	-	C
Hungary	C	C	C	C	-	-	C
Italy	C	C	C	C	-	-	R
Luxembourg	C	C	C	R	-	-	-
Norway	C	C	C	R	-	-	R
Poland	C	C	C	C	-	-	R
Portugal	C	C	C	C	C	-	-
Spain	C	C	C	C	-	-	C
Sweden	C	C	C	R	-	-	-
Switzerland	C	C	C	C	-	-	-
United Kingdom	-	-	C	-	-	-	-
Yugoslavia	C	C	C	C	-	C	C

This Sale £ 004·38
Litres 009·98
044·2
Four Star
Unleaded
UNLEADED

Owning a car is always going to involve some expense, and the running costs can generally be divided into two main areas. The first concerns fixed costs which cannot be avoided, such as car tax and insurance (although obviously you can shop around for the best insurance quote). However, for most owners savings can be made in the second main area of expense, which covers fuel costs and servicing/maintenance bills.

It's surprisingly easy to reduce the amount of fuel used, simply by adapting your driving style to suit the prevailing conditions, and avoiding certain driving habits which tend to increase fuel consumption unnecessarily.

A large proportion of the average servicing bill is made up of garage labour time, money which can be saved by carrying out the work yourself. Refer to *'Servicing'* on page 80 for easy-to-follow instructions on how to carry out most servicing work. It's also worth bearing in mind that maintenance costs can be cut significantly by reducing unnecessary wear-and-tear on the car.

The following advice deals with the most significant causes of high fuel consumption and unnecessary wear-and-tear, and explains how to save money and reduce environmental pollution during everyday driving.

- **Don't warm the engine up with the car standing still** – Engine wear and pollution is at its highest when the engine is warming up, and the engine takes a long time to warm up when running at idle speed with the car stationary. To avoid excessive wear and pollution, drive off as soon as the engine starts, and don't use more 'revs' than necessary.
- **On cars with a manual choke, don't use the choke any longer than necessary** – Push the choke control fully in as soon as the engine will run smoothly without it. When the engine is running with the choke applied, extra fuel is being used and so the fuel consumption is increased, and more pollution is produced.
- **Avoid sudden full throttle acceleration** – Sudden acceleration increases fuel consumption, engine wear and pollution.
- **Don't drive at high engine speeds** – Minimum fuel consumption and pollution is achieved at low engine speeds and in the highest possible gear. Lower engine speed also means less noise and engine wear. For maximum economy, stay in as high a gear as possible for as long as possible, without making the engine labour.
- **Don't always drive at maximum speed** – Fuel consumption, pollution and noise increase rapidly at high speeds. A small reduction in speed (particularly during motorway driving) can significantly lower fuel consumption and pollution.
- **Look well ahead, and drive as smoothly as possible** – By looking well ahead you will be able to react to any change in road conditions in plenty of time, allowing you to brake and accelerate smoothly. Unnecessary or harsh acceleration and braking increases fuel consumption and pollution.
- **Where possible, avoid dense, slow-moving traffic** – In these conditions, more frequent braking, acceleration and gear changing are required, which increase fuel consumption and pollution.
- **Stop the engine during traffic hold-ups** – Obviously if your engine has stopped it doesn't use any fuel and there is no pollution.

- **Check the tyre pressures regularly** – Low tyre pressures increase the rolling resistance of the car, and therefore increase fuel consumption, as well as increasing tyre wear and causing handling problems.
- **Don't carry unnecessary luggage** – Weight has a significant effect on fuel consumption, especially in dense traffic where frequent acceleration is required.
- **Don't leave a roof rack fitted when not in use** – The extra air resistance increases fuel consumption.
- **Switch off any unnecessary electrical circuits as soon as possible** – Heated rear windows, foglights, heater blowers, etc, consume a considerable amount of electrical power. The engine must work harder to provide this power by driving the alternator, and the fuel consumption is therefore increased.
- **Check the fuel consumption regularly** – By doing this you will be able to notice any significant increase in fuel consumption, and any problem which may be causing it can be investigated before it develops into anything more serious.
- **Ensure that your car is serviced regularly** – This will ensure that the car operates as efficiently as possible, reducing fuel consumption and pollution.
- **If your car engine is suitable, use unleaded petrol** – This will reduce pollution. If in any doubt as to whether your car's engine is suitable for use with unleaded petrol, seek advice from the car's manufacturer, or from a recognised dealer.

Crime against cars is a serious problem, and many owners suffer the consequences of theft or damage every year. The likelihood of your car becoming a target for criminals can be reduced by making it more difficult for your car to be broken into or stolen, and the following advice should prove helpful. Some of the points may seem obvious, but the majority of cars are broken into or stolen in a very short space of time with little force or effort required.

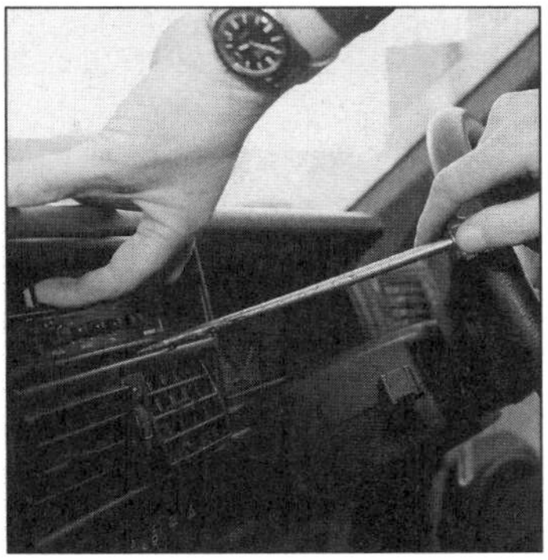

- **Always remove the ignition key** – even in a garage or driveway at home. Always make sure that the steering column lock is engaged after removing the key.
- **Always lock your car** – even in a garage or driveway at home. Where fitted, ensure that 'deadlocks' are engaged before leaving the car and, where applicable, don't forget to lock the fuel filler cover. Make sure that all the windows and the sunroof or folding roof, where applicable, are properly shut. If the car is in a garage, lock the garage. If your car is stolen or broken into while it's unlocked, your insurance company may not pay for the full value of loss or damage.
- **Never leave valuable items on display** – Even if you're only leaving your car for a few minutes, move anything that might be attractive to a thief (even a coat or briefcase) out of sight, and preferably take it with you or lock it safely in the boot. Don't leave valuable items (especially credit cards) in the glovebox. Don't leave your vehicle documents in the car (registration document, MOT certificate, insurance certificate, etc), as they could help a thief to sell it.
- **Park in a visible and (preferably) busy area** – This will deter thieves as they run a greater risk of being caught. If you're parking your car at night, try to leave it in a well-lit area.
- **Put your radio aerial down (where applicable) when parking** – Radio aerials can prove an attractive target for vandals.
- **Protect any in-car entertainment equipment** – The latest security-coded equipment won't work if someone tampers with it and disconnects it from the battery. Some equipment is specially designed so that you can remove the control panel or the entire unit, and take it with you when you leave the car.
- **Fit lockable wheel nuts (or bolts, as applicable)** – if your car is fitted with expensive alloy wheels. Alloy wheels are a favourite target for thieves.
- **Have your car windows etched with the registration number** – This will help to trace your car if it's stolen and the thieves try to change its identity. Other glass components such as sunroofs and headlamps can also be etched if desired. Many garages and specialists can provide this service, and it's possible to buy DIY glass etching kits from motor accessory shops if you prefer to tackle the job yourself.
- **Fit a vehicle immobiliser device** – Many different types are available, but the most common types consist of a substantial metal bar which can be locked in place between the steering wheel and the pedals to prevent the car from being driven.
- **Have an alarm fitted** – Many different types are available, and some are expensive, but they will deter thieves. Some alarms have built-in immobiliser devices to prevent the car from being driven. If you have an alarm fitted, remember to switch it on even if you're only leaving the car for a few minutes.

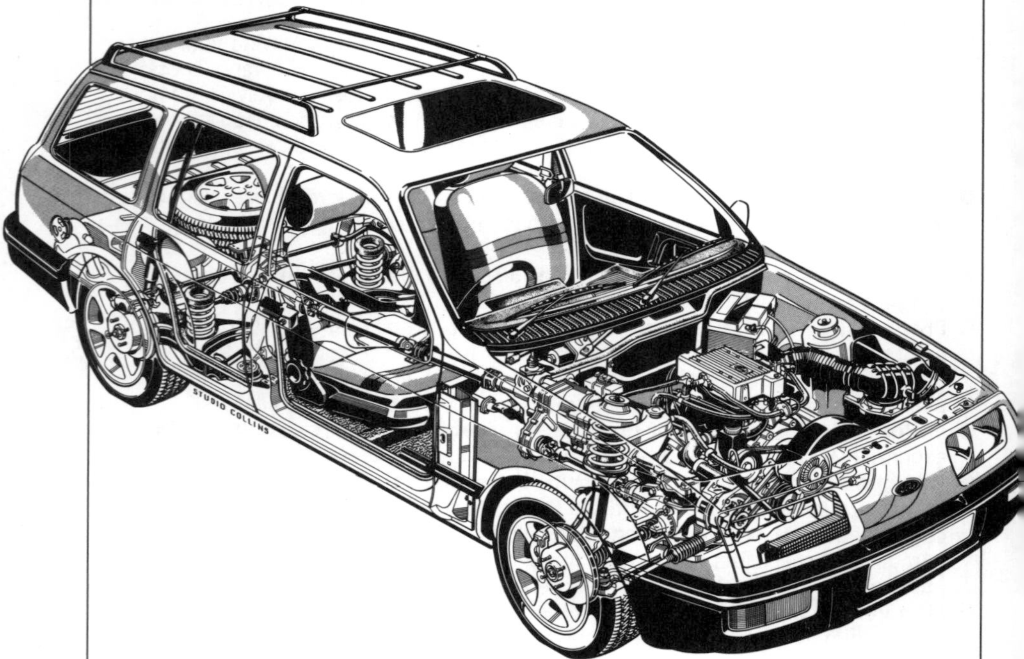
STUDIO COLLINS

Lubricant and fluid capacities [Litres (Pints)]

Fuel tank capacity

All models	**60**	(13.2 gallons)

Engine oil

Quantity of oil required to bring level on dipstick from 'MIN' to 'MAX' mark	**1.00**	(1.8)

Engine oil capacity (for oil change)

OHC engines	**3.75**	(6.6)
DOHC engines	**4.50**	(7.9)
CVH engines	**3.50**	(6.2)
V6 engines (two-wheel-drive models)	**4.25**	(7.5)
V6 engines (four-wheel-drive models)	**4.65**	(8.2)

Coolant capacity (for coolant change)

OHC engines	**8.00**	(14.1)
DOHC carburettor engines	**7.90**	(13.9)
DOHC fuel injection	**7.30**	(12.8)
CVH engines	**9.50**	(16.7)
V6 engines	**8.50**	(15.0)

Spark plugs

Note: *Consult a Ford dealer for the latest 'Motorcraft' spark plug recommendations.*

Type	Champion
OHC engines	**RF7YC**
DOHC engines	**RC7YCC**
CVH engines	**RC7YCC**
V6 engines:	
2.3 and 2.8 litre	**RN7YCC**
2.9 litre non-catalyst	**RC7YCC**
2.9 litre catalyst	**RS9YCC**

Electrode gap [mm (inches)]

OHC engines	**0.70**	(0.028)
DOHC engines	**0.80**	(0.032)
CVH non-catalyst engines	**0.80**	(0.032)
CVH catalyst engines	**1.00**	(0.039)
V6 engines except 2.9 litre catalyst	**0.80**	(0.032)
V6 2.9 litre catalyst engine	**1.00**	(0.039)

Tyre pressures – cold [bars (lbf/in²)]

Note: *The following is intended as a guide only. Manufacturers frequently change tyre pressure recommendations, and it is suggested that a Ford dealer is consulted for latest recommendations.*

	Front	Rear
Up to 3 occupants (all models)	**1.8** (26)	**1.8** (26)
Fully laden:		
Saloon (Sapphire) and Hatchback models	**2.0** (29)	**2.5** (36)
Estate models up to 1987:		
175 R 13-S, 175 R 13-T and 175 R 13-H tyres	**2.0** (29)	**2.8** (40)
196/60 R 14-H and 195/60 VR 14 tyres	**2.0** (29)	**2.5** (36)
195/70 R 13-H tyres	**2.0** (29)	**3.3** (48)
Estate models from 1987:		
All except 195/60 VR 14 tyres	**2.0** (29)	**3.3** (48)
195/60 VR 14 tyres	**2.0** (29)	**2.5** (36)

Recommended lubricants and fluids

Component/system	Lubricant/fluid type and specification	Duckhams recommendation
1 ENGINE	Multigrade engine oil, viscosity range SAE 10W/30 to 20W/50, to API SG/CD or better	Duckhams QXR, QS, Hypergrade Plus or Hypergrade
2 COOLING SYSTEM	55% clean water and 45% antifreeze to Ford specification SSM-97B9103-A	Duckhams Universal Antifreeze and Summer Coolant
3 BRAKING SYSTEM	Brake fluid to Ford specification SAM-6C9103-A	Duckhams Universal Brake and Clutch Fluid
MANUAL GEARBOX	Refer to Owners Workshop Manual	
4 AUTOMATIC TRANSMISSION	Automatic transmission fluid to Ford specification SQM-2C9010-A	Duckhams Uni-Matic
5 POWER STEERING SYSTEM	Automatic transmission fluid to Ford specification SQM-2C9010-A	Duckhams Uni-Matic

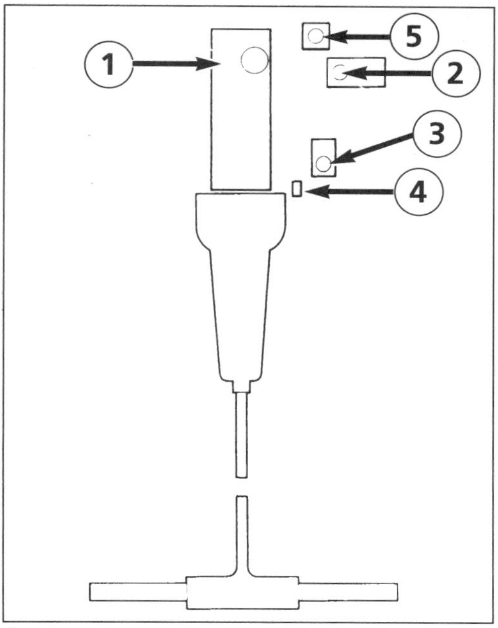

▲ *OHC engine models*

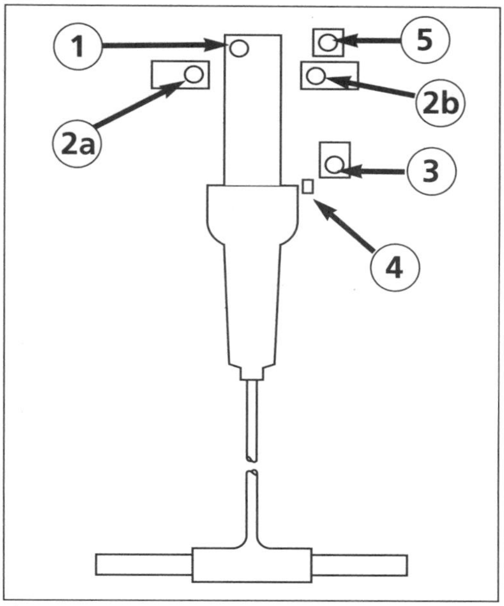

▲ *DOHC engine models*
2a *Carburettor models* ***2b*** *Fuel injection models*

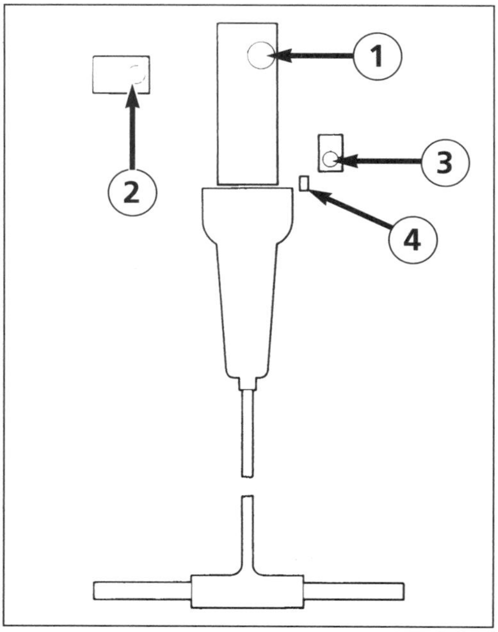

▲ *CVH engine models*

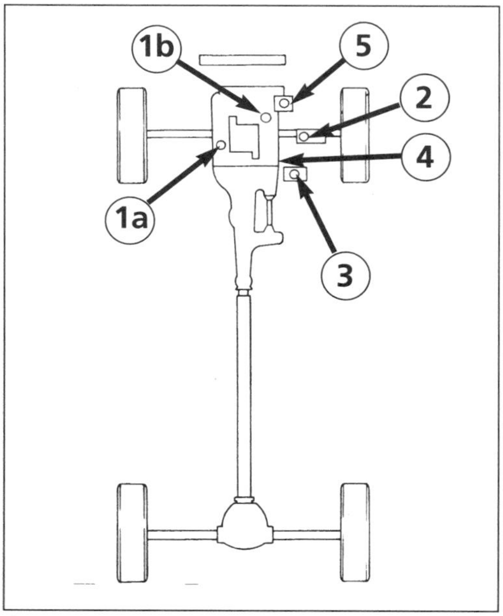

▲ *V6 engine models*
1a *2.8 litre engine* ***1b*** *2.9 litre engine*

376
08
EU·R

REGULAR CHECKS

To ensure that your car is reliable and safe to drive, there are one or two essential checks which are so simple that they're often ignored. These checks only take a few minutes, and could save you a lot of inconvenience and expense. It's a good idea to carry out these checks once a week, and certainly before you start off on a long journey. This Chapter explains how to carry out the checks, and what to do if things aren't quite as they should be.

Whenever you're carrying out checks or servicing jobs, safety must always be the first consideration, and you should bear in mind the advice given in the *'Safety first!'* notes on page 86 before proceeding.

CHECKS

The following checks should be carried out regularly. The checks are explained in more detail in the following pages.

- *Check the oil level*
- *Check the coolant level*
- *Check the brake fluid level*
- *Check the tyres*
- *Check the washer fluid level*
- *Check the battery electrolyte level (where applicable)*
- *Check the wipers and washers*
- *Check the lights and horn*
- *Check for fluid leaks*

ITEMS REQUIRED WHEN CARRYING OUT CHECKS

When carrying out the regular checks, it's a good idea to have the following items close-to-hand. You won't always need all the items, but it's as well to have them available just in case.

● *Small quantity of clean rag* ● *Tyre pressure gauge and foot pump* ● *Small screwdriver (for removing stones from tyres)* ● *Pin (for adjusting washer nozzles)* ● *1.0 litre pack of engine oil (of the correct type – refer to **'Service specifications'** on page 68)* ● *Small quantity of coolant solution (made up from approximately 55% clean water and 45% antifreeze)* ● *Small container of brake fluid (must be an airtight container)* ● *Washer fluid additive* ● *Small quantity of distilled water – for models fitted with batteries which require topping-up*

***Note:** For details of how to identify engine types, refer to **'Servicing'** on page 89.*

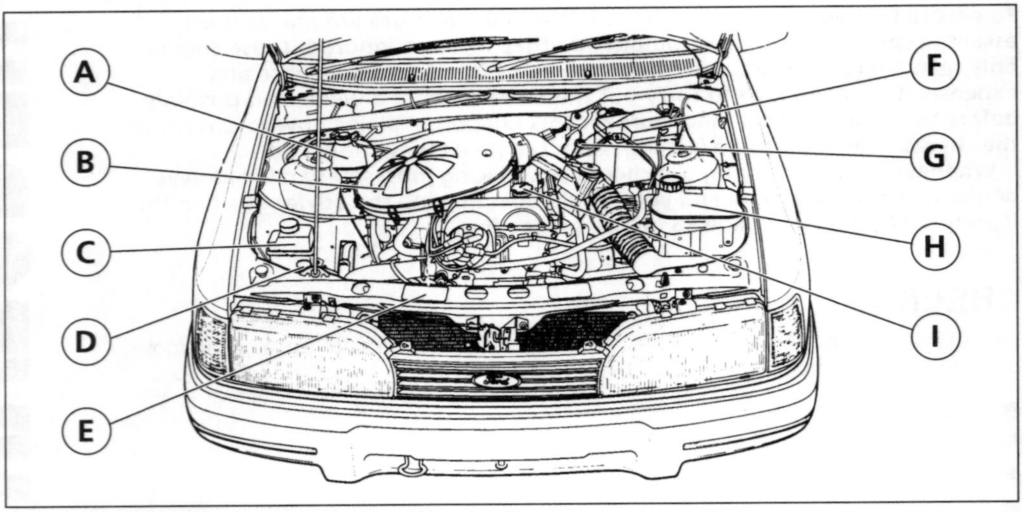

▲ *Typical engine compartment – DOHC engine with carburettor*

A *Brake fluid reservoir*
B *Air cleaner*
C *Power steering fluid reservoir*
D *Windscreen washer fluid reservoir*
E *Vehicle identification plate*
F *Battery*
G *Engine oil dipstick*
H *Engine coolant reservoir*
I *Engine oil filler cap*

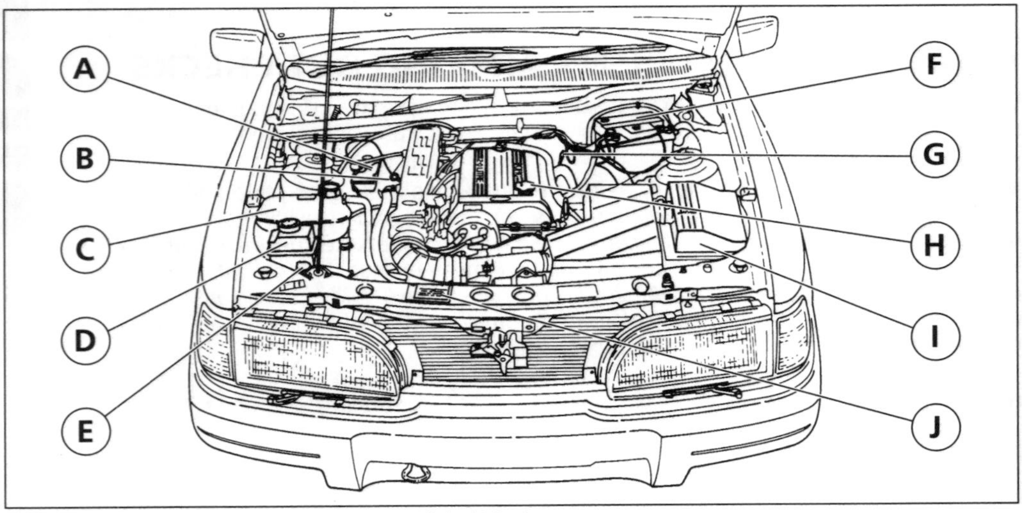

▲ *Typical engine compartment – DOHC engine with fuel injection*

A *Brake fluid reservoir*
B *Automatic transmission fluid dipstick*
C *Engine coolant reservoir*
D *Power steering fluid reservoir*
E *Windscreen washer fluid reservoir*
F *Battery*
G *Engine oil dipstick*
H *Engine oil filler cap*
I *Air cleaner*
J *Vehicle identification plate*

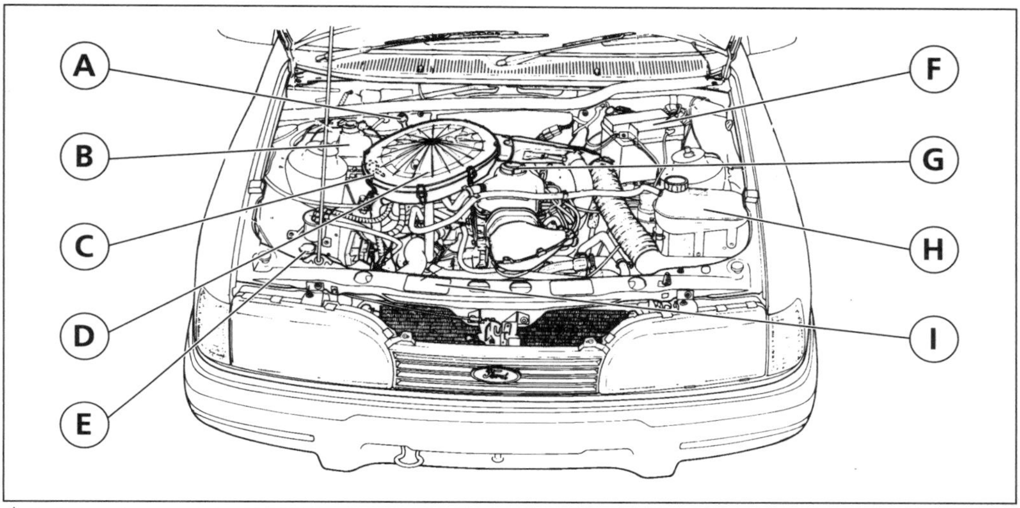

▲ *Typical engine compartment – CVH engine*

- ***A*** *Engine oil dipstick*
- ***B*** *Brake fluid reservoir*
- ***C*** *Automatic transmission fluid dipstick*
- ***D*** *Air cleaner*
- ***E*** *Windscreen washer fluid reservoir*
- ***F*** *Battery*
- ***G*** *Engine oil filler cap*
- ***H*** *Engine coolant reservoir*
- ***I*** *Vehicle identification plate*

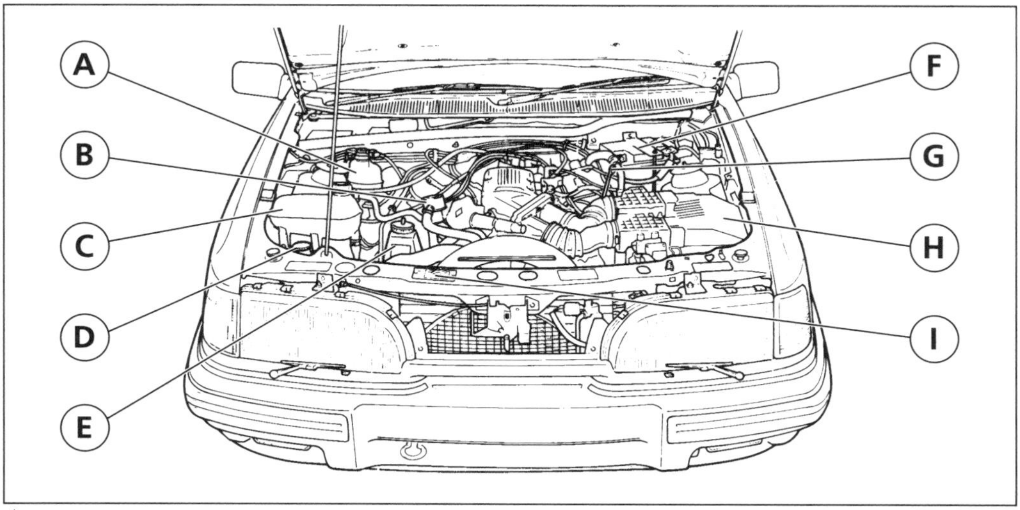

▲ *Typical engine compartment – 2.9 litre V6 engine*

- ***A*** *Brake fluid reservoir*
- ***B*** *Engine oil filler cap*
- ***C*** *Engine coolant reservoir*
- ***D*** *Windscreen washer fluid reservoir*
- ***E*** *Power steering fluid reservoir*
- ***F*** *Battery*
- ***G*** *Engine oil dipstick*
- ***H*** *Air cleaner*
- ***I*** *Vehicle identification plate*

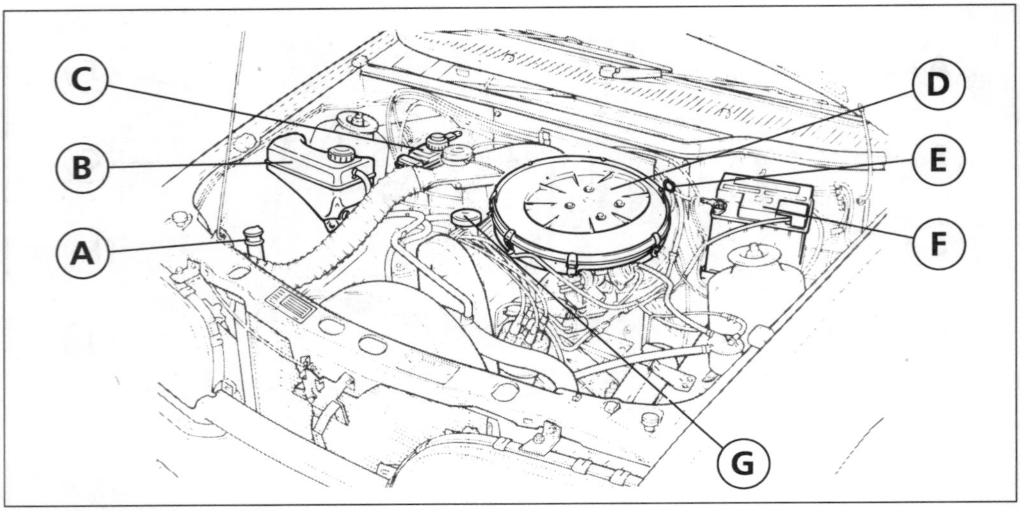

▲ *Typical engine compartment – OHC engine with carburettor*

A *Washer fluid reservoir*
B *Coolant expansion tank*
C *Brake fluid reservoir*
D *Air cleaner*
E *Engine oil dipstick*
F *Battery*
G *Engine oil filler cap*

Checking oil level

To check the oil level, the car should be standing on level ground with the engine switched off. If the engine has just been run, wait for a few minutes to allow the level to settle before taking a reading.

Open the bonnet, and look for the oil

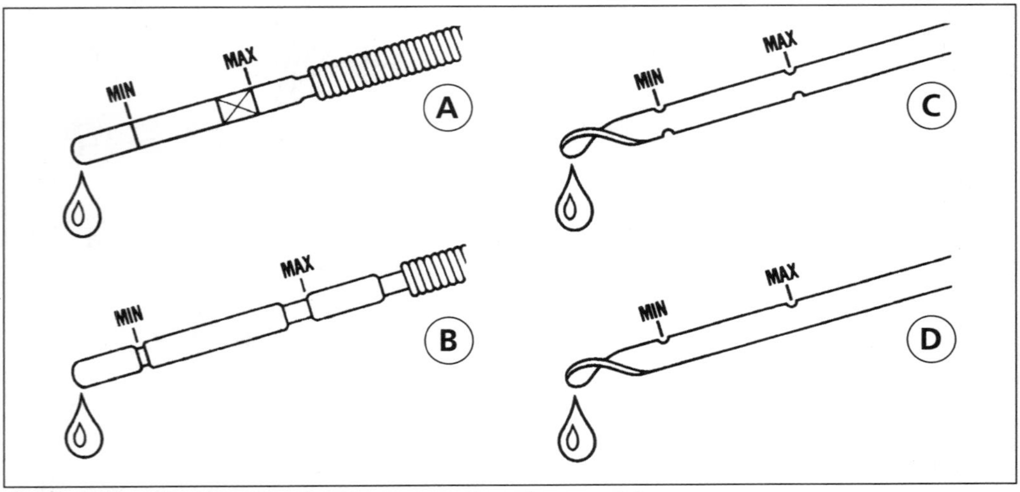

▲ *Engine oil level dipstick markings (markings may vary depending on model)*

A *CVH engine*
B *Later OHC engine*
C *Later V6 engine*
D *DOHC engine*

dipstick. The location of the oil dipstick varies depending on model – refer to the accompanying illustrations for details.

Pull out the dipstick and wipe it clean (with a clean non-fluffy cloth). Slowly push it back into its tube, and then pull it out again. On some models, an oil level sensor is fitted to the dipstick so, where applicable, take care not to damage the wires when moving the dipstick. Note the oil level, which should never be allowed to drop below the 'MIN' mark.

If the oil needs topping-up, pull the oil filler cap from the top of the engine (or from the tube at the side of the engine on V6 engine models), and top-up to the 'MAX' mark on the dipstick. Note that the amount of oil required to raise the level on the dipstick from the 'MIN' to the 'MAX' mark is approximately 1.0 litre (1.75 pints). Always try to use the same type and make of oil, and take care not to overfill. Mop up any oil which might have been spilt, and make sure that the oil filler cap is correctly refitted.

If the oil needs to be topped-up regularly, check for leaks, and if necessary seek advice.

▲ *Topping-up the engine oil level – OHC engine with carburettor*

Checking coolant level

The coolant level can be checked visually by checking the level in the expansion tank when the engine is **cold**. The level should be between the 'MIN' and 'MAX' marks on the side of the tank.

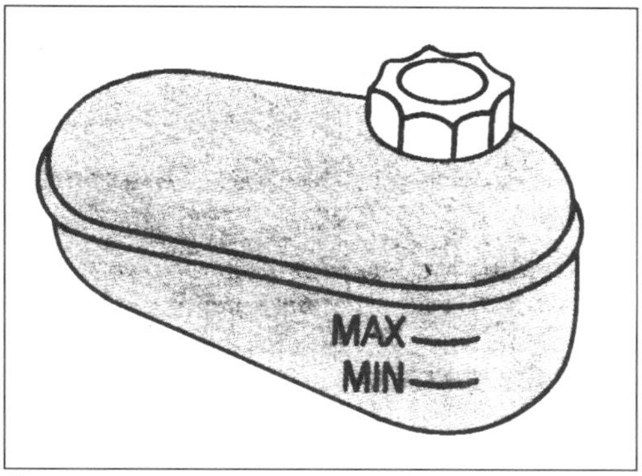

▲ *Coolant level markings on expansion tank*

If topping-up is necessary, first the cooling system pressure cap must be removed.

Topping-up should always be done with the engine cold, but if it does prove necessary to remove the cooling system pressure cap with the engine hot for any reason, take care to avoid scalding. Place a thick rag over the cap, and loosen the cap slowly in stages to gradually release the pressure in the system.

A screw-type pressure cap is fitted to the expansion tank. Slowly unscrew the cap a quarter of a turn, wait for the pressure in the system to be released, then remove the cap.

Coolant should be added through the expansion tank. Topping-up should always be done with a mixture of water and antifreeze to the same strength as the mixture already in the system. Although plain water can be used, it's

▲ *Topping-up the coolant level – OHC engine model*

unwise to make it a habit, as the strength of the antifreeze in the main system will gradually be diluted. If there's any doubt about the strength of the coolant mixture, it's better to add too much antifreeze than not enough.

Normally, topping-up will rarely be required, and if the need for regular topping-up arises, it will probably be due to a leak somewhere in the system. Leaks are most likely to occur from the radiator or the various hoses. If no leaks can be found, it's possible that there's an internal fault in the engine, such as a crack in the cylinder head, or a blown cylinder head gasket, but in this case it's best to seek specialist advice. It's important to note that because of the different types of metals used in the engine, it's vital to use antifreeze with suitable anti-corrosion additives all year round. **Never** use plain water to fill the whole system.

Checking brake fluid level

Note: *Refer to **'Safety first'** on page 86 for special precautions which should be taken when handling brake fluid.*

Although a low brake fluid level warning light is fitted, the fluid level should be checked visually whenever the oil and coolant levels are checked. The level should be up to the 'MAX' mark on the side of the reservoir, but it will fall slightly as the brake friction material wears. The fluid level must **never** be allowed to drop below the 'MIN' mark.

▲ *Topping-up the brake fluid level – CVH engine model*

If topping-up is required, always use the correct type of fluid, which should always be stored in a full, air-tight container. Don't top-up using fluid which has been stored in a partly-full container, as it will have absorbed moisture from the air, which can reduce its performance.

Topping-up should hardly ever be required unless there's a leak somewhere in the hydraulic system. If a leak is suspected, the car should not be driven until the braking system has been thoroughly checked. **Never** take any risks where brakes are concerned.

Checking tyres

It's extremely important to carry out regular checks on the tyres, making sure that the pressures are correct, and that the tyres are not damaged. The tyres are the only part of the car in contact with the road, and so their condition will affect the steering and general handling of the car, and therefore its safety.

To check tyre pressures accurately, the tyres must be cold, which means that the car must not have been driven recently. Note that it can make a noticeable difference to the tyre pressures if the car has been standing out in the sun. This can be very noticeable when one side of the car is in the sun, and the other side is in shadow.

Note that the recommended tyre pressures vary depending on the size of tyres fitted – refer to *'Service specifications'* on page 68. The size of the tyre (eg 155 SR 13) is clearly marked on the tyre sidewall, although the style of the tyre size marking may vary depending on the tyre manufacturer – consult a Ford dealer or a tyre specialist for further details of tyre size markings and the latest pressure recommendations.

If the tyres are being checked after the car has been driven, for example when filling up with petrol during a journey, the pressures are bound to be higher than specified. It's best to simply check that the pressures in the two front tyres are *equal*, and similarly for the two back tyres (remember that the specified pressure for the rear tyres may be different to the front). **Never** let air out of the tyres of a car which has recently been driven to bring the pressures down to that specified (unless the tyre pressures have been increased for a fully-loaded car, and the load has just been reduced).

When checking the tyre pressures, don't forget to check the spare!

With the tyres properly inflated, run your fingers around the edge of the tyre to check for any cuts and bulges in the tyre walls. Ideally, the car should be jacked up to visually check the tyres all round. Alternatively, the car can be moved backwards and forwards to enable you to check all round the tread. Check all round the tread for cuts, and remove any debris, such as small stones and broken glass, using a screwdriver or similar tool. Any large items such as nails which have penetrated the surface of the tyre should be left in the tread (to identify the location of the damage) until the tyre has been repaired. Refer to *'Breakdowns'* on page 44 for details of how to fit the spare wheel.

Also check the tyre treads for wear. Uneven wear across one particular tyre may be due to a fault with the suspension or steering. By law, the tread depth must be at least 1.6 mm (0.06 in) throughout a continuous band comprising the central three-quarters of the width of the tyre tread around the full circumference of the tyre.

If you find that any of the tyres is excessively worn or damaged, obtain a new tyre as soon as possible. It's a good idea to try and stick to one type of tyre if possible, rather than mixing several different makes on the car. Generally, you'll find that it's much cheaper to buy tyres from a tyre specialist, rather than an ordinary garage. Make sure that when you buy a new tyre you have the wheel balanced, otherwise you might find that the new tyre causes vibration when driving.

Shoulder wear

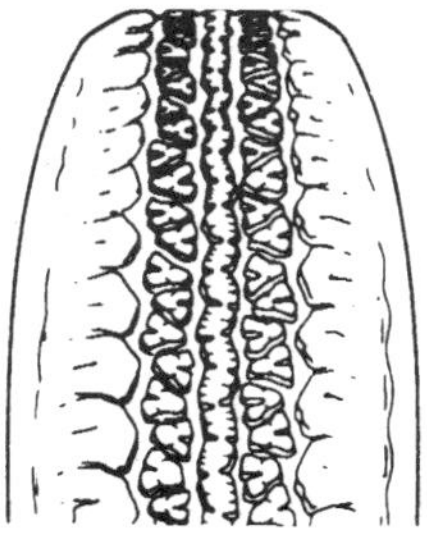

Probable cause:
Underinflation (wear on both sides)
Action: Check and adjust pressure

Probable cause:
Incorrect wheel camber (wear on one side)
Action: Repair or renew suspension parts

Probable cause:
Hard cornering
Action: Reduce speed

Centre wear

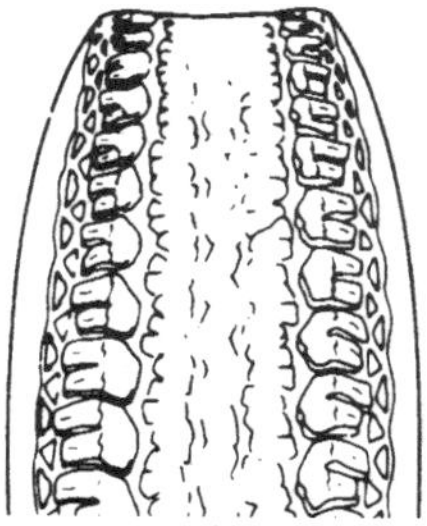

Probable cause:
Overinflation
Action: Measure and adjust pressure

Uneven wear

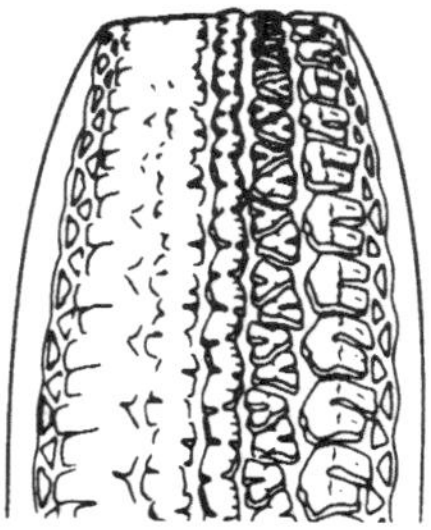

Probable cause:
Incorrect camber or castor
Action: Repair or renew suspension parts

Probable cause:
Malfunctioning suspension
Action: Repair or renew suspension parts

Probable cause
Unbalanced wheel
Action: Balance tyres

Probable cause:
Out-of-round brake disc/drum
Action: Machine or renew disc/drum

Toe wear

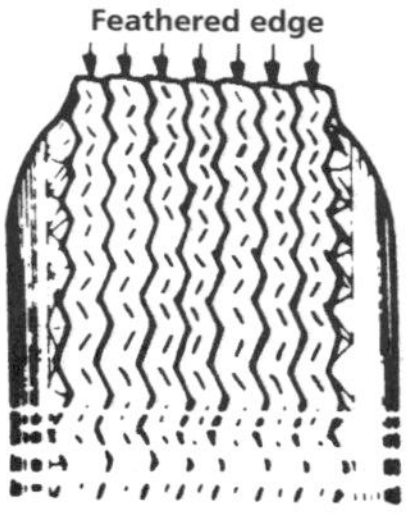

Probable cause:
Incorrect toe setting
Action: Adjust front wheel alignment

▲ *Tyre wear patterns and causes*

Checking washer fluid level

The windscreen washer fluid reservoir is located at the front right-hand side of the engine compartment (when viewed from the driver's seat).

On early models with a tailgate washer system, the tailgate washer fluid reservoir is located behind a panel at the left-hand side of the luggage compartment.

On later models, the tailgate washer and windscreen washer systems share the same reservoir in the engine compartment.

On models fitted with headlamp washers, the headlamp washer system shares the same reservoir as the windscreen washer system in the engine compartment.

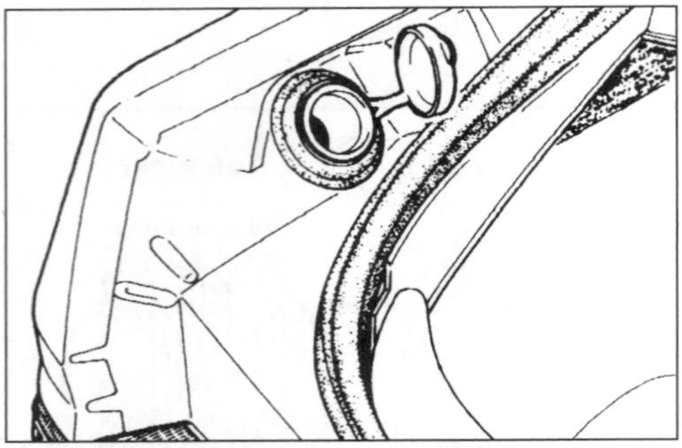

▲ *Tailgate washer fluid reservoir location – early models*

If topping-up is necessary, pull the cap from the reservoir, and top-up as necessary with clean water. A suitable washer fluid additive will keep the glass free from smears, and will prevent the fluid freezing in winter.

Checking battery electrolyte level

From approximately 1982, Ford cars have been fitted with a maintenance-free battery during production. The maintenance-free battery is 'sealed for life', and does not require topping-up with distilled water. A maintenance-free battery is usually clearly marked, and there will be no removable covers to enable topping-up.

On early models with a standard or low-maintenance type battery (or on later models where the maintenance-free battery has been replaced by one of this type from another source), the electrolyte level should be checked as follows.

The battery is located at the rear left-hand side of the engine compartment (when viewed from the driver's seat). Remove the caps or the cover from the top of the battery, and check the level of the electrolyte fluid (with some types of battery you can see the fluid level through the battery case). The level should be just above the tops of the metal plates inside the battery, or in some cases, up to a mark on the battery case.

If necessary, add distilled water (**not** ordinary tap water) to each cell, to bring the level just above the tops of the battery plates, or up to the marks, as applicable. With some types of battery, distilled water is added to a trough in the top of the battery until all the filling slots are full, and the bottom of the trough is just covered. On completion, refit the caps or cover, and carefully wipe up any drops of water that were spilt.

Topping-up should hardly ever be required, and the need for frequent topping-up indicates that the battery is being overcharged due to a fault in the charging circuit – seek advice from someone suitably qualified if this is the case.

Checking wipers and washers

Check all the wipers and washers to make sure that they're working properly. Don't allow the wipers to work on dry glass for too long, as it will strain the motor and wear out the wiper blades. If necessary, the washer nozzles can be adjusted using a pin inserted into the end of the nozzle. If you're adjusting the windscreen washer nozzles, remember to aim them fairly high on the windscreen, as the airflow will usually deflect the spray down when the car is moving.

Over a period of time, the wiping action of the wiper blades will deteriorate, causing smearing of the glass. The rubber may also crack, particularly at the edges of the blades. When this happens, the blades must be renewed. This is a straightforward job, and is accomplished as follows:

Pull the wiper arm away from the glass, against the spring pressure, until the arm clicks into position. Press the small retaining clip on the blade, then slide the blade out of the

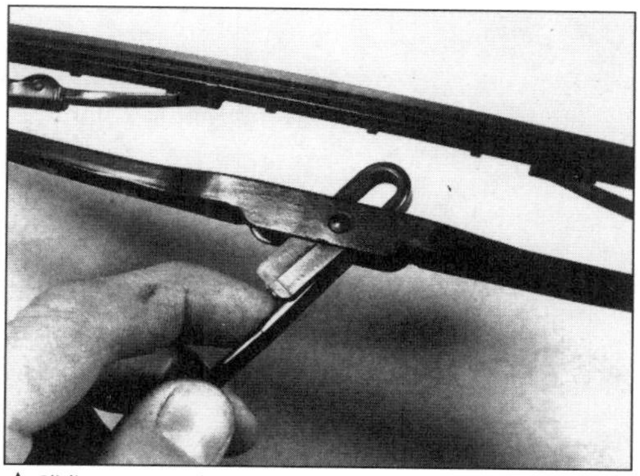

▲ *Sliding a wiper blade from the arm*

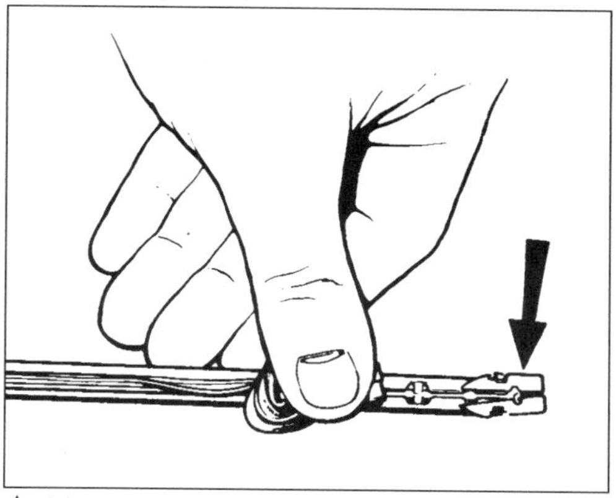

▲ *Sliding back the wiper blade rubber to expose the spring clip (arrowed)*

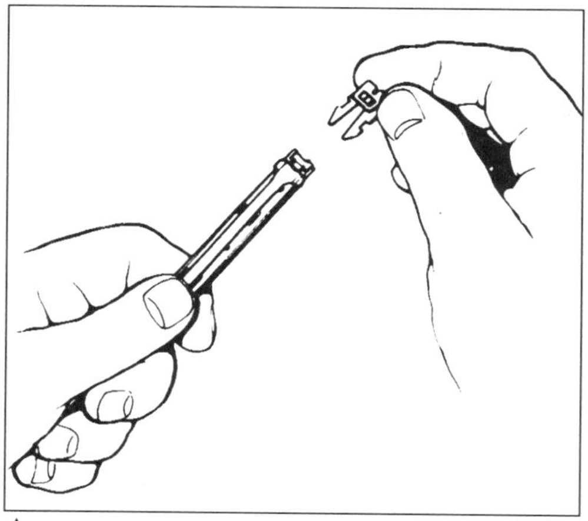

▲ *Removing the spring clip from the wiper blade*

hooked part of the arm. The blade can be renewed complete, or the rubber insert can be renewed as follows:

The rubber is held in the arm by a metal spring clip at one end. Slide back the rubber from the end of the blade, then squeeze the spring clip and pull it from the blade. The rubber can now be slid from the end of the blade. Fit the new rubber and/or the blade by reversing the removal operations.

Checking lights and horn

Switch on all the lights and check that they're working. Don't forget to check the direction indicators and the brake lights (the ignition must be switched on to check these), either with the help of an assistant, or by looking for the reflection in a suitable window or door etc.

Check that the direction indicators work with the brake lights on, and that the brake lights work with the tail lights on. Some faults may cause the various rear lights to interact, which can be dangerous, as it may confuse or distract following drivers.

Check the operation of the horn. A short 'blast' should be enough to prove that it at least works, but from time to time check its operation over several seconds.

If any of the bulbs need renewing, refer to *'Bulb, fuse and relay renewal'* on page 119. Any other problems are likely to be caused by a faulty switch, or loose or corroded connections (especially earth connections), and it's best to seek advice to solve these.

Checking for fluid leaks

Check the ground where the car is normally parked for any stains or spots of fluid which may have been caused by fluid leaking from the car.

Open the bonnet, and make a quick check of all the hoses and pipes, and the surfaces of all the components in the engine compartment. If there's any sign of leaking fluid, try to find the source of the leak, and seek advice if necessary. It can be difficult to identify leaking fluids, but if there's obviously a major leak, or if you suspect even a slight brake fluid or petrol leak, don't drive the car until the problem has been investigated by someone suitably qualified.

SERVICING

Regular servicing will ensure that your car is reliable and safe to drive, and could save you a lot of money in the long run. Many of the unexpected expenses which can crop up if things go wrong with your car can be avoided by carrying out regular servicing.

A significant proportion of the average garage servicing bill (well over 50% in many cases) is made up of labour costs, so obviously a lot of money can be saved by carrying out the work yourself. You'll find that most of the servicing jobs are very straightforward, and you don't need to be a mechanical genius to 'have a go' yourself. Don't be put off when you look under your bonnet; there are surprisingly few items which require frequent attention, and most of those which do are easily accessible without the need for anything more than basic tools. You'll discover that DIY servicing can be very rewarding, and will help you to understand what makes your car 'tick'.

The following chart lists all the servicing tasks recommended by the car manufacturer, and details of how to carry out the necessary work can be found in the subsequent pages. A few of the jobs require more extensive knowledge, or the use of special tools, and detailed explanation is beyond the scope of this Handbook. These more complicated tasks are identified on the chart, and details can be found in our Owners Workshop Manual for your particular model. Even if you decide to have the work done by a garage, the chart will enable you to check that the necessary work has been carried out.

Whenever you're carrying out servicing, safety must always be the first consideration, and you should read through the *'Safety first!'* notes on page 86 before proceeding any further.

ENGINE COMPARTMENT

Check all components for fluid leaks, corrosion or deterioration

Clean/check air conditioning condenser

ENGINE

Check engine oil level

Renew engine oil and filter

Remove and clean oil filler cap

On OHC and V6 engines, renew crankcase emission valve

On OHC and V6 engines, check valve clearances

On V6 engines, check inlet manifold bolts

COOLING SYSTEM

Check coolant level

On OHC engines, check coolant pump/alternator drivebelt

FUEL AND EXHAUST SYSTEMS

Check/lubricate throttle linkage

Renew air cleaner filter element

On carburettor/1.6CVH engine models, check air cleaner air intake temperature control

Check/adjust idle speed and mixture settings

On fuel injection models, renew fuel filter

IGNITION SYSTEM

Renew spark plugs

Clean distributor cap (where applicable), coil and HT leads

MANUAL GEARBOX

Check gearbox oil level

On four-wheel-drive models, check transfer box oil level

AUTOMATIC TRANSMISSION

Check transmission fluid level

Check gear selector and kickdown linkages

Check brake band adjustment

FINAL DRIVE/DIFFERENTIAL

Check final drive/differential oil level

Note: *In addition to the tasks listed in the table, carry out the following:*
Every 36 000 miles*, on OHC and CVH engines, renew the timing belt (optional on OHC engines, compulsory on CVH engines) – refer to Owners Workshop Manual.*
Every 36 000 miles or 2 years*, renew the coolant – refer to* ***'Service tasks'*** *on page 112.*

Regular Weekly	6000 miles (10 000 km) 6 months	12 000 miles (20 000 km) 1 year	24 000 miles (40 000 km) 2 years	36 000 miles (60 000 km) 3 years
■	■	■	■	■
		□	□	□
■	■	■	■	■
	■	■	■	■
	■	■	■	■
			■	
		□	□	□
		□	□	□
■	■	■	■	■
		■	■	■
	■	■	■	■
			■	
				■
	□	□	□	□
			□	
		■	■	■
			■	
		□	□	□
		□	□	□
		■	■	■
		□	□	□
			□	
		□	□	□

Items marked □ are considered to be beyond the scope of this Handbook, and details can be found in the relevant Owners Workshop Manual for your car (OWM 903 for Sierra 4-cylinder engine models, and OWM 904 for Sierra V6 engine models).

BRAKING SYSTEM

Check brake fluid level

Check brake fluid level warning light

Check handbrake

Check front disc pad thickness

Check rear brake pad/shoe thickness

Renew brake hydraulic fluid

SUSPENSION AND STEERING

Check tyres for condition and pressure

Check tightness of roadwheel bolts

Check power steering fluid level

Check power steering pump drivebelt

Check all components for damage or wear

INTERIOR AND BODYWORK

Check seat belts

Inspect paintwork and bodywork

Lubricate locks and hinges

Check operation of locks and windows

ELECTRICAL SYSTEM

Check washer fluid level

Check battery electrolyte level (where applicable)

Check wipers and washers

Check operation of all lights and electrical equipment

Check alternator drivebelt

CAR UNDERSIDE

Check all components for fluid leaks, corrosion or deterioration

ROAD TEST

Check instruments and electrical equipment

Check steering, suspension and general handling

Check engine, clutch (where applicable), gearbox and driveshafts

Check braking system

At the time of writing, no full service schedule was available for 1.6 litre CVH engine models, and it is suggested that a Ford dealer is consulted for the latest service recommendations on these vehicles.

Regular Weekly	6000 miles (10 000 km) 6 months	12 000 miles (20 000 km) 1 year	24 000 miles (40 000 km) 2 years	36 000 miles (60 000 km) 3 years
■	■	■	■	■
	■	■	■	■
		■	■	■
	□	□	□	□
	□	□	□	□
				□
■	■	■	■	■
	■	■	■	■
	■	■	■	■
			■	
	□	□	□	□
	■	■	■	■
	■	■	■	■
		■	■	■
		■	■	■
■	■	■	■	■
■	■	■	■	■
■	■	■	■	■
■	■	■	■	■
		■	■	■
	■	■	■	■
		■	■	■
		■	■	■
		■	■	■
		■	■	■

Items marked □ are considered to be beyond the scope of this Handbook, and details can be found in the relevant Owners Workshop Manual for your car (OWM 903 for Sierra 4-cylinder engine models, and OWM 904 for Sierra V6 engine models).

SAFETY FIRST!

Professional motor mechanics are trained in safe working procedures. No matter how enthusiastic you may be about getting on with the job you have planned, take time to read this Section. Don't risk an injury by failing to follow the simple safety rules explained here. The following is a guide to encourage you to be 'safety conscious' as you work on your car.

Essential points

ALWAYS use a safe system of supporting your car when working underneath it. The car's own jack (or a single hydraulic jack) is never sufficient. Once you've raised the car you should use a safe means of holding it up. Axle stands, ramps or substantial wooden blocks are good; concrete blocks and bricks may crack or disintegrate and should not be used. Make sure that you locate the supports where you know the car won't collapse and where they can't slip.

ALWAYS loosen wheel nuts and other 'high torque' nuts or bolts (as applicable for the task to be carried out) before jacking up your car, or it may slip and fall.

ALWAYS make sure that your car is in **Neutral** or in **Park** (in the case of automatic transmission), and that the handbrake is securely on before trying to start the engine.

UNLESS the engine is cold, **ALWAYS** cover the cooling system's pressure cap (expansion tank cap) with several thicknesses of cloth before trying to remove it. Release the pressure slowly or the coolant may escape suddenly and scald you.

ALWAYS wait until the engine has cooled down before draining the engine (and/or transmission) oil, or the coolant. Oil or coolant that is very hot may scald you. If you can touch the engine sump/transmission/radiator (as applicable) without discomfort, the engine has probably cooled sufficiently to avoid scalding.

ALWAYS allow the engine to cool before working on it. Many parts of the engine become very hot in normal operation and you may burn yourself badly.

ALWAYS keep brake fluid, antifreeze and other similar liquids away from your car's paintwork as it may damage the finish (or remove the paint altogether).

ALWAYS keep toxic fluids, such as fuel, brake and transmission fluids and antifreeze off your skin, and never syphon them by mouth.

ALWAYS wear a mask when doing any dusty work, or where spraying is involved, especially when doing body repair and brake jobs.

ALWAYS clean up oil and grease spills – someone may slip and be injured.

ALWAYS use the right tool for the job. Spanners and screwdrivers which don't fit properly are likely to slip and cause injury.

ALWAYS get help to lift and handle heavy parts – it's never worth risking an injury.

ALWAYS take time over the jobs you take on. Plan out what you must do, and make sure that you have the right tools and spare parts. Follow the recommended steps, and check over the job once you're finished. Leave 'short-cuts' to the experts.

ALWAYS wear eye protection when using power tools such as drills, sanders and grinders, when using a hammer, and when working beneath your car.

ALWAYS use a barrier cream on your hands when doing dirty jobs, especially when in contact with fuel, oils, greases, and brake and transmission fluids. It will help protect your skin against infection and will make the dirt easier to remove later. Make sure that your hands aren't slippery. The use of a suitable specialist hand cleaner will make dirt and grease easier to remove without causing infection or damage to skin. Long term or regular contact with used oils and fuel can be a health hazard.

ALWAYS keep loose clothing and long hair out of the way of any moving parts.

ALWAYS take off rings, watches, bracelets and neck chains before starting work on your car, especially the electrical system.

ALWAYS make sure that any jacking equipment or lifting tackle you use has a safe working load which will cope with what you intend to do, and use the equipment exactly as recommended by the manufacturers.

ALWAYS keep your work area tidy; someone may trip or slip on articles left lying around and be injured.

ALWAYS get someone to check up on you from time to time if you're working on the car alone, to make sure that you're alright.

ALWAYS do the work in a logical order, and check that you've put things back together properly and that everything is tightened as it should be.

ALWAYS keep children and animals away from an unattended car, and from the area where you're working.

ALWAYS park cars with catalytic converters away from materials which may burn, such as dry grass, oily rags, etc if the engine has recently been run. Catalytic converters reach extremely high temperatures, and any such materials close by may catch fire.

REMEMBER that your safety, and that of your car and other people rests with you. If you are in any doubt about anything, get specialist advice and help right away.

IF in spite of these precautions you injure yourself – get medical help immediately.

Asbestos

Some parts of your car, such as brake pads and linings, clutch linings and gaskets contain asbestos and you should take appropriate precautions to avoid inhaling dust (which is hazardous to health) when working with them. If in doubt, assume that they **do** contain asbestos.

Fire

Remember at all times that petrol is highly flammable. Never smoke, or have any kind of naked flame around, when working on the car. But the risk doesn't end there – a spark caused by an electrical short-circuit, by two metal surfaces contacting each other, by careless use of tools, or even by static electricity built up in your body under certain conditions, can ignite petrol vapour, which in a confined space is highly explosive.

If a fuel leak is suspected, try to find the cause and seek advice as soon as possible, and never risk fuel leaking onto a hot engine or exhaust. Catalytic converters (where fitted) run at extremely high temperatures, and therefore can be an additional fire hazard – observe the precautions outlined previously in this Section.

It's recommended that a fire extinguisher of a type suitable for fuel and electrical fires is kept handy in the garage or workplace at all times.

Never try to extinguish a fuel or electrical fire with water.

Fumes

Certain fumes are highly toxic and can quickly cause unconsciousness and even death if inhaled to any extent, especially if inhalation takes place through a lighted cigarette or pipe. Petrol vapour comes into this category, as do the vapours from certain solvents such as trichloroethylene. Any draining or pouring of such fluids should be done in a well ventilated area.

When using cleaning fluids and solvents, read the instructions carefully. Never use materials from unmarked containers – they may give off poisonous vapours.

Never run a car engine in an enclosed space such as a garage. Exhaust fumes contain carbon monoxide which is extremely poisonous; if you need to run the engine, always do so in the open air or at least have the rear of the car outside the workplace. Although cars fitted with catalytic converters produce far less toxic exhaust gases, the above precautions should still be observed.

If you're fortunate enough to have the use of an inspection pit, never run the engine while the car is standing over it; the fumes, being heavier than air, will concentrate in the pit with possibly lethal results.

The battery

Never short across the two poles of the battery! (A battery is shorted by connecting the two terminals directly to each other, and this can happen accidentally when working under the bonnet with metal tools.) The heavy discharge caused will create 'gassing' of hydrogen from the battery and this is highly explosive. Shorting across a battery may also cause sparks and the combination can cause the battery to explode with potentially lethal results. A conventional battery will normally be giving off a certain amount of hydrogen all the time, so for the same reasons given above, never cause a spark or allow a naked flame close to it.

Batteries which are sealed for life require special precautions which are normally outlined on a label attached to the battery. Such precautions usually relate to battery charging and jump starting from another vehicle (for details of jump starting refer to *'Breakdowns'* on page 47).

If possible, loosen the filler plugs or cover when charging the battery from an external source. Don't charge at an excessive rate or the battery may burst. Special care should be taken with the use of high charge-rate boost chargers to prevent the battery from overheating.

Take care when topping-up and when carrying the battery. The battery contains dilute sulphuric acid which is very corrosive. It will burn and may cause long-term damage if in contact with skin, eyes or clothes. Similarly, corrosive deposits around the battery terminals may be harmful.

Always wear eye protection when cleaning the battery to prevent the corrosive deposits from entering your eyes.

Mains electricity and electrical equipment

When using any electrical equipment which works from the mains, such as an electric drill or inspection light, always ensure that the appliance is correctly connected to its plug and that, where necessary, it's properly earthed. Always use an RCD (Residual Current Device, a safety device incorporating a circuit breaker) when using mains electrical equipment. Don't use such appliances in damp conditions, and take care not to create a spark or apply excessive heat close to fuel or fuel vapour. Also make sure that the appliances meet the relevant national safety standards.

Ignition HT voltage

A severe electric shock can result from touching certain parts of the ignition system, such as the spark plug HT leads, when the engine is running or being cranked, particularly if components are damp or the insulation is faulty. Where an electronic ignition system is fitted, the HT voltage is much higher than that used in a contact breaker system, and could prove fatal, especially to wearers of cardiac pacemakers.

Disposing of used engine oil

Used engine oil is a hazard to health and the environment, and should be disposed of safely and cleanly. Most local authorities provide a disposal site which will have a special tank for waste oil. If in doubt, contact your local council for advice on where you can dispose of oil safely – there may be a local garage who will allow you to use their specialised oil disposal tank free of charge. Remember that it's not just the oil which causes a problem, but empty containers, old oil filters, and oily rags, so these should be taken to your local disposal site too.

Do not under any circumstances pour engine oil down a drain, or bury it in the ground.

Jacking and vehicle support

The jack provided with the car is designed for emergency wheel changing, and should not be used for servicing and overhaul work. Instead, a more substantial workshop jack (trolley jack or similar) should be used. Whichever type is used, it's essential that additional safety support is provided by means of axle stands designed for this purpose. Never use makeshift means such as narrow wooden blocks or piles of house bricks, as these can easily topple or, in the case of bricks, disintegrate under the weight of the car. Further information on the correct positioning of the jack and axle stands is provided at the end of this Section.

If you don't need to remove the wheels, the use of drive-on ramps is recommended. Ensure that the ramps are correctly aligned with the wheels, and that the car is not driven too far along them so that it promptly falls off the other end or tips the ramps.

JACK AND AXLE STAND POSITIONS

When using a trolley jack or any other type of workshop jack to raise the front of the car, the jack head should be placed under the centre of the front lower crossmember, with a shaped block of wood between the jack head and the crossmember to act as an insulator.

To raise the rear of the car, the jack head should be placed under the centre of the differential casing, with a shaped block of wood between the jack head and the crossmember to act as an insulator.

Do not jack the car under the engine sump or any of the steering or suspension components.

Axle stands can be positioned under the jacking points provided for use with the car jack, or under the points already described for use with a trolley jack.

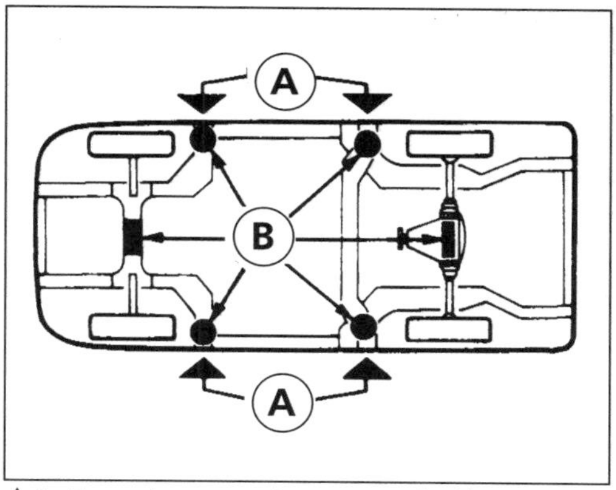

▲ *Jacking and support points on car underside*
A *Jacking points for use with wheel-changing jack*
B *Jacking points for use with trolley jack or axle stands*

BUYING SPARE PARTS

When buying spare parts such as an oil filter, or an air filter, make sure that the correct replacement parts are obtained for your particular car. Many changes are made during the production run of any car, and when ordering spare parts, it will usually be necessary to know the year the car was built, the model type (eg 'L' or 'Ghia') and engine size, and in some cases the Vehicle Identification Number (VIN). The VIN is stamped on a metal plate attached to the body panel above the radiator in the engine compartment, and usually appears on the Vehicle Registration Document.

All of the components required for servicing will be available from an official Ford dealer, but most parts should also be available from good motor accessory shops and motor factors.

Identifying engine types

Four different types of engine have been fitted to the Sierra models covered by this Handbook, which can be identified as follows:

OHC (OVERHEAD CAMSHAFT) ENGINES

OHC engines have been produced in 1.3, 1.6, 1.8 & 2.0 litre capacities, and have a distributor cap which is located at an angle, at the front left-hand side (when viewed from the driver's seat) of the engine.

DOHC (DOUBLE OVERHEAD CAMSHAFT)

DOHC engines are all of 2.0 litre capacity, and have 'DOHC 2.0 LITRE' stamped on the camshaft cover on top of the engine. The distributor cap protrudes from the front of the engine on the right-hand side (when viewed from the driver's seat).

CVH (COMPOUND VALVE ANGLE, HEMISPHERICAL COMBUSTION CHAMBERS) ENGINES

CVH engines have been produced in both 1.6 and 1.8 litre capacities, and have an alternator mounted on the left-hand side (when viewed from the driver's seat).

V6 ENGINES

V6 engines have been produced in 2.3, 2.8 & 2.9 litre capacities, and have two banks of three cylinders mounted in a 'V' formation. The distributor cap is mounted vertically, at the back of the engine.

▲ *Underbonnet component locations on a 1983 2.0 litre OHC carburettor engine model (air cleaner removed for clarity)*

- **1** *Fusebox*
- **2** *Windscreen wiper motor*
- **3** *Battery earth terminal*
- **4** *Ignition coil*
- **5** *Carburettor*
- **6** *Distributor cap*
- **7** *Fuel pressure regulator*
- **8** *Radiator top hose*
- **9** *Cooling system thermostat housing*
- **10** *Radiator cooling fan shroud*
- **11** *Vehicle Identification Number (VIN) plate*
- **12** *Alternator*
- **13** *Number 1 spark plug HT lead*
- **14** *Engine oil filler cap*
- **15** *Windscreen washer fluid reservoir*
- **16** *Cooling system expansion tank pressure cap*
- **17** *Brake fluid reservoir*

▲ *Underbonnet component locations on a 1985 2.0 litre OHC fuel injection engine model*

1 *Battery earth terminal*
2 *Brake servo vacuum valve*
3 *Ignition coil*
4 *Fuel filter*
5 *Air cleaner*
6 *Fuel pressure regulator (fuel injection system)*
7 *Distributor cap*
8 *Airflow meter (fuel injection system)*
9 *Radiator cooling fan*
10 *Throttle housing*
11 *Alternator*
12 *Vehicle Identification Number (VIN) plate*
13 *Windscreen washer fluid reservoir*
14 *Cooling system expansion tank pressure cap*
15 *Number 1 spark plug HT lead*
16 *Engine oil filler cap*
17 *Inlet manifold*
18 *Brake fluid reservoir*
19 *Fusebox*
20 *Windscreen wiper motor*
21 *Engine oil level dipstick*

▲ *Underbonnet component locations on a 1989 1.8 litre CVH engine model*

1 *Battery earth terminal*
2 *Ignition coil*
3 *Cooling system expansion tank pressure cap*
4 *Number 1 spark plug HT lead*
5 *Alternator*
6 *Distributor cap shroud*
7 *Vehicle Identification Number (VIN) plate*
8 *Radiator cooling fan*
9 *Radiator top hose*
10 *Washer fluid reservoir*
11 *Fuel vapour separator*
12 *Cooling system thermostat housing*
13 *Oil filler cap*
14 *Carburettor*
15 *Brake fluid reservoir*
16 *Fusebox*
17 *Windscreen wiper motor*
18 *Engine oil level dipstick*
19 *Camshaft cover*

▲ *Underbonnet component locations on a 1985 2.8 litre V6 engine model*

- ***1*** *Fusebox*
- ***2*** *Windscreen wiper motor*
- ***3*** *Distributor*
- ***4*** *Engine oil level dipstick*
- ***5*** *Battery earth terminal*
- ***6*** *Engine oil filler cap*
- ***7*** *Ignition coil*
- ***8*** *Plenum chamber (fuel injection system)*
- ***9*** *Accelerator cable*
- ***10*** *Fuel distributor (fuel injection system)*
- ***11*** *Air cleaner*
- ***12*** *Fuel injector pipes*
- ***13*** *Radiator bottom hose*
- ***14*** *Cooling system thermostat housing*
- ***15*** *Radiator cooling fan*
- ***16*** *Alternator/power steering pump drivebelts*
- ***17*** *Radiator top hose*
- ***18*** *Vehicle Identification Number (VIN) plate*
- ***19*** *Power steering fluid reservoir*
- ***20*** *Windscreen washer fluid reservoir*
- ***21*** *Cooling system expansion tank pressure cap*
- ***22*** *Brake fluid reservoir (ABS)*

▲ *Underbonnet component locations on a 1990 2.9 litre V6 engine model*

1 *Fusebox*
2 *Windscreen wiper motor*
3 *Distributor*
4 *Accelerator cable*
5 *Battery earth terminal*
6 *ABS hydraulic valve block*
7 *Ignition coil*
8 *Engine oil level dipstick*
9 *Air cleaner*
10 *Airflow meters (fuel injection system)*
11 *Throttle housing*
12 *Plenum chamber (fuel injection system)*
13 *Radiator cooling fan*
14 *Alternator/power steering pump drivebelts*
15 *Vehicle Identification Number (VIN) plate*
16 *Radiator top hose*
17 *Windscreen washer fluid reservoir filler*
18 *Power steering fluid reservoir*
19 *Cooling system expansion tank pressure cap*
20 *Engine oil filler cap*
21 *Brake fluid reservoir (ABS)*

SERVICE TASKS

The following pages provide easy-to-follow instructions to enable you to carry out the regular service tasks recommended by the manufacturer.

Every 250 miles (400 km) or weekly

Refer to *'Regular checks'* on page 70 and carry out all the checks described.

Every 6000 miles (10 000 km) or 6 months – whichever comes first

ENGINE COMPARTMENT

Check all hoses and pipes, and the surfaces of all components for signs of fluid leakage, corrosion or deterioration

If there's any sign of leaking fluid, try to find the source of the leak and seek advice if necessary. It can be difficult to identify leaking fluids, but if there's obviously a major leak, or if you suspect even a slight brake fluid or petrol leak, don't drive the car until the problem has been investigated by someone suitably qualified.

Check all rubber or fabric hoses for signs of cracking, damage due to rubbing on other components, and general deterioration. It's sensible to have any suspect hoses renewed as a precaution against possible failure.

Check for obvious signs of corrosion on metal pipes, hose clips, and component joints, which may indicate a fluid leak. Any badly-corroded components should be cleaned and, if necessary, renewed. Pay particular attention to metal fluid pipes, and have them renewed if they're badly pitted or corroded.

ENGINE

Renew the engine oil and filter

Note: *The following items will be required for this task:*

- *A suitable quantity of engine oil of the correct type – refer to* ***'Service specifications'*** *on page 68.*
- *New oil filter (of the correct type for your particular car)*
- *New oil drain plug sealing washer (depending on the condition of the old one)*
- *Suitable container to catch the old oil as it drains (make sure that the capacity of the container is larger than the oil capacity of the engine – refer to* ***'Service specifications'*** *on page 66).*
- *Oil filter wrench – for removal of old oil filter*
- *Small quantity of clean rag*
- *Suitable spanner or socket to fit oil drain plug*

The oil should be drained when the engine is warm, immediately after a run. Ideally, the front of the car should be jacked up to improve access. If the front of the car is to be raised, apply the handbrake, then jack up the front of the car and support it securely on axle stands (refer to 'Jacking and vehicle support' in *'Safety first!'* on page 88 for details of where to position the jack and axle stands).

Position a suitable container under the engine sump to catch all the oil which will be drained.

Remove the oil filler cap from the top of the engine, then working under the car, unscrew the drain plug from the engine sump using a suitable spanner or a socket. The drain plug is located towards the rear of the left-hand side (as viewed from the driver's seat) of the sump on all except CVH engines, and towards the rear of the right-hand side (as viewed from the driver's seat) of the sump on CVH engines. Oil will be released before the drain plug is removed completely, so take precautions against scalding, as the oil will be hot. Position the container to allow for the fact that the initial jet of oil may fall at least eight inches away from the drain hole, and try not to drop the drain plug when removing it.

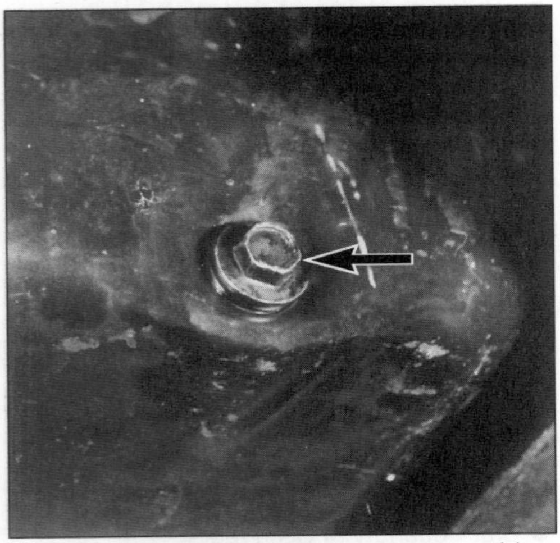

▲ *Engine oil drain plug (arrowed) – OHC engine model*

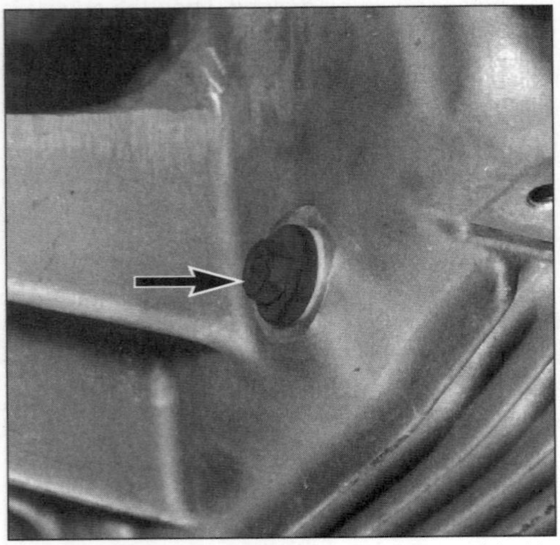

▲ *Engine oil drain plug (arrowed) – V6 engine model with four-wheel-drive*

Allow the oil to drain for at least 15 minutes, moving the container slightly if necessary from its original position.

Check the condition of the oil drain plug sealing washer, and renew it if necessary.

When the oil has finished draining, clean the drain plug, washer, and the mating face of the sump, then refit and tighten the drain plug.

Position a suitable container under the oil filter. The oil filter is located on the left-hand side (as viewed from the driver's seat) of the cylinder block on OHC and DOHC engines, and towards the rear of the right-hand side (as viewed from the driver's seat) of the cylinder block on CVH and V6 engines. On some cars, access to the oil filter may be easier from underneath the car.

Unscrew the filter using a suitable chain or strap wrench if necessary to loosen it.

Wipe clean the oil filter mounting flange on the engine. Smear a little clean engine oil on the sealing ring of the new oil filter, then screw the filter onto the engine, and tighten it *by hand only*. If no instructions are provided with the filter, tighten it until the sealing ring touches the mounting face on the engine, then tighten it a further three-quarters of a turn.

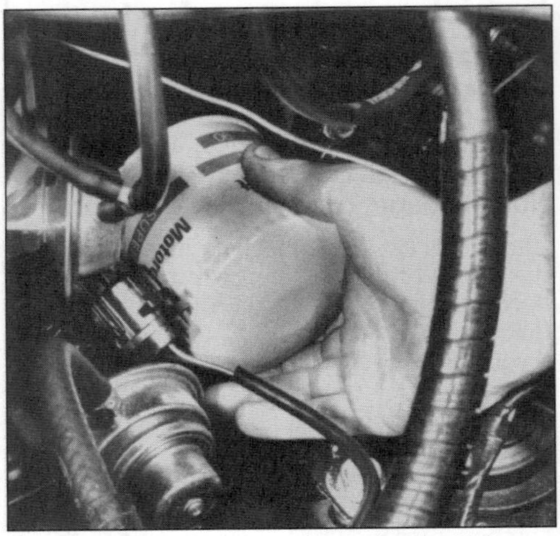

▲ *Fitting a new oil filter (viewed from above) – OHC engine*

Where applicable, lower the car to the ground and, with the car parked on level ground, fill the engine with the correct quantity and grade of oil through the filler on top of the engine. Fill the engine slowly over a period of several minutes, until the level reaches the **MAX** mark on the dipstick.

Make sure that the oil filler cap is fitted on completion.

Don't race the engine when starting it for the first time after an oil change, as the oil will take a few seconds to circulate. There may be a delay before the oil pressure warning light goes out, as the engine lubrication system fills with oil.

Run the engine and check for leaks from the filter and the drain plug, then stop the engine and check the oil level (refer to *'Regular checks'* on page 74 for details). It is likely that a small quantity of oil will need to be added.

Dispose of the old engine oil safely (refer to *'Safety first!'* on page 88). **Don't** pour it down a drain.

Remove and clean the oil filler cap

Remove the oil filler cap from the top of the engine.

Carefully clean the inside of the cap, taking care not to damage the filter gauze, where applicable (the cap can be cleaned using paraffin, but dry it thoroughly before refitting).

Where applicable, check the condition of the O-ring seal at the bottom of the filler cap, and renew it if it's damaged or worn.

COOLING SYSTEM

Check the coolant level, and top-up if necessary

Refer to *'Regular checks'* on page 75.

FUEL AND EXHAUST SYSTEMS

Check the operation of the throttle linkage, and lubricate if necessary

Move the accelerator pedal through its full range of travel, and check that the throttle operates smoothly, without any stiffness.

Any stiffness may be due to a lack of lubrication at the pedal pivot, or at the throttle mechanism under the bonnet.

If necessary, apply a *few drops* of light oil to the pedal pivot in the driver's footwell, and/or to the throttle linkage under the bonnet.

To identify the throttle linkage, ask an assistant to work the accelerator pedal while you look under the bonnet. It should be possible to see the accelerator cable moving the throttle mechanism on the carburettor or the throttle housing (in the case of fuel injection models) as the pedal is moved. Note that on some carburettor models, it may be necessary to remove the air cleaner for access to the throttle linkage – refer to the procedure for air cleaner filter element renewal in the *'24 000 mile (40 000 km)'* service Section on page 104 for details.

BRAKING SYSTEM

Check the brake fluid level, and top-up if necessary

Refer to *'Regular checks'* on page 76.

Check the operation of the brake fluid level warning light

Open the bonnet, then unscrew the filler cap from the top of the brake hydraulic fluid reservoir. Allow the float under the filler cap to hang downwards (if necessary disconnect the wiring from the cap to enable the cap to be removed, but reconnect the wiring once the cap has been removed). Hold a piece of rag or a cloth under the filler cap to collect any drips of hydraulic fluid – don't let hydraulic fluid drip onto the paintwork.

▲ *Brake fluid reservoir cap float (arrowed)*

Chock the wheels to prevent the car from moving, then switch on the ignition and release the handbrake.

The handbrake 'on'/low brake fluid level warning light should stay on when the handbrake is released, and assuming that the fluid level is not low, the warning light should then go out when the filler cap is refitted to the fluid reservoir (with the handbrake still released).

If the warning light doesn't operate as described, it's likely that the circuit is faulty, and the problem should be investigated by someone suitably qualified.

Re-apply the handbrake and remove the chocks from the wheels on completion.

Refit the filler cap using a reversal of the removal procedure.

SUSPENSION AND STEERING

Check the tyres for damage, tread depth and uneven wear

Refer to *'Regular checks'* on page 76.

Check and if necessary adjust the tyre pressures

Refer to *'Regular checks'* on page 76.

Check the tightness of the roadwheel bolts

Using the wheel brace (refer to *'Changing a wheel'* on page 44), check that the bolts cannot be turned easily using reasonable force. Do not overtighten the bolts.

Check the power steering fluid level, and top-up if necessary (where applicable)

All except DOHC engine models

The power steering fluid reservoir is located at the front left- or right-hand side of the engine – the reservoir is mounted on the power steering fluid pump, and can be identified from the drivebelt(s) running to the pump pulley directly in front of the reservoir.

The fluid level is indicated by a dipstick attached to the reservoir filler cap.

Ensure that the ignition is switched off.

Remove the filler cap/dipstick by turning anti-clockwise, then wipe the dipstick clean (with a clean non-fluffy cloth) and slowly refit the filler cap.

Remove the filler cap/dipstick again, and note the fluid level.

There are three different types of dipstick markings: **HOT** and **FULL COLD**, **H** and **C**, or **MAX** and **MIN**.

If the fluid is at its normal operating temperature (ie the car has recently been driven), the fluid level should be up to the **HOT**, **H**, or **MAX** mark. If necessary, add the specified type of fluid (refer to 'Service specifications' on page 66) to bring the level up to the relevant mark.

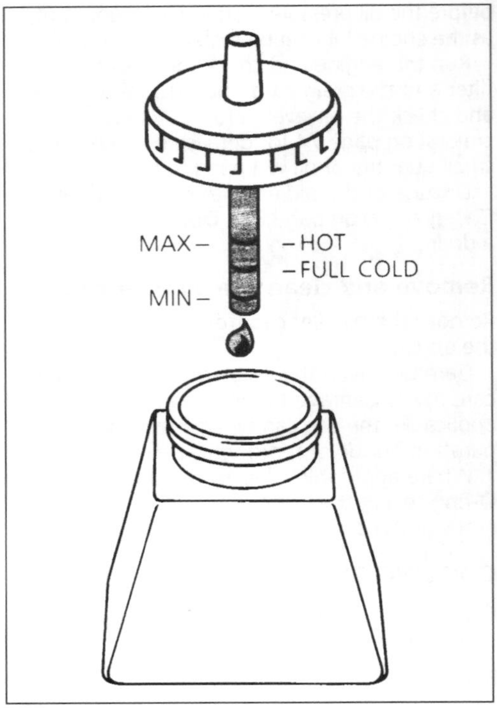

▲ *Power steering fluid level dipstick markings – all later models except DOHC engine models*

If the fluid is cold, the fluid level should be up to the **FULL COLD** or **C** mark, and above the **MIN** mark. If necessary, add the specified type of fluid to bring the level up to the relevant mark, but note that the level should be rechecked with the fluid at normal operating temperature, as described previously.

Make sure that the filler cap is correctly refitted and tightened on completion.

DOHC engine models

The power steering fluid reservoir is mounted at the front right-hand (when viewed from the driver's seat) corner of the engine compartment.

The fluid level is indicated by **MAX** and **MIN** marks on the side of the fluid reservoir.

If the fluid is at its normal operating

temperature (ie the car has recently been driven), the fluid level should be up to the **MAX** mark. If necessary, ensure that the ignition is switched off, then remove the filler cap and add the specified type of fluid (refer to *'Service specifications'* on page 68) to bring the level up to the mark.

If the fluid is cold, the fluid level should be up to the **MIN** mark. If necessary, add the specified type of fluid to bring the level up to the mark, but note that the level should be rechecked with the fluid at normal operating temperature, as described previously.

Make sure that the filler cap is correctly refitted and tightened on completion.

INTERIOR AND BODYWORK

Check the seat belt webbing for cuts and damage, and check the seat belt operation

If any of the seat belt webbing is damaged, frayed or worn, the relevant seat belt(s) must be renewed.

Check that all inertia reel seat belts can be pulled smoothly from their retractors, and that they return smoothly, unassisted.

Check that the locking mechanism works correctly by sharply tugging the seat belt webbing.

If a seat belt retractor or locking mechanism appears to be faulty, the relevant seat belt(s) should be renewed.

Carefully inspect the paintwork for damage, and the bodywork for corrosion

Touch up any minor stone chips, and have any corroded areas repaired before the damage becomes serious.

For details of how to treat minor paint scratches, refer to *'Bodywork and interior care'* on page 118.

ELECTRICAL SYSTEM

Check the washer fluid level(s) and top-up if necessary

Refer to *'Regular checks'* on page 78.

Check and if necessary top-up the battery electrolyte level (where applicable)

Refer to *'Regular checks'* on page 78.

Check the operation of the windscreen wipers and washers

Refer to *'Regular checks'* on page 78.

Check the operation of all lights, electrical equipment and accessories

Refer to *'Fault finding'* on page 131 if any problems are discovered.

If any of the bulbs need to be renewed, refer to *'Bulb, fuse and relay renewal'* on page 119.

CAR UNDERSIDE

Check all hoses and pipes, and the surfaces of all components for signs of fluid leakage, corrosion, damage or deterioration

Refer to *'Engine compartment'* checks at the beginning of the *'6000 miles (10 000 km)'* service Section on page 95.

In addition, check the exhaust system and the fuel tank for any sign of damage or serious corrosion (light surface rust is to be expected).

Check the exhaust mountings to make sure that they're secure.

Check the visible suspension and steering components for obvious signs of damage or wear.

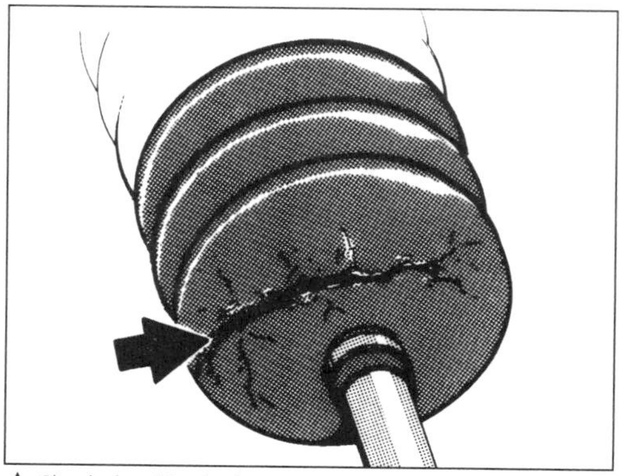

▲ *Check the driveshaft and steering gear rubber gaiters for splits (arrowed)*

Check the driveshafts for obvious signs of distortion or damage.

Pay particular attention to the driveshaft and steering gear rubber gaiters, which should be renewed if they're split, or if there are any obvious signs of lubricant leakage.

ADDITIONAL TASKS

Note that as well as the tasks described in the preceding paragraphs, the additional tasks given in the *'Service schedule'* chart on page 82 should be carried out every 6000 miles (10 000 km) or 6 months – whichever comes first. These tasks require more detailed explanation, or the use of special tools, and are considered beyond the scope of this Handbook. For details of these additional tasks, refer to the Owners Workshop Manual.

Every 12 000 miles (20 000 km) or 12 months – whichever comes first

In addition to all the items in the *'6000 miles (10 000 km)'* service on page 95, carry out the following.

COOLING SYSTEM

On OHC engine models, check the condition and adjustment of the coolant pump/alternator drivebelt

Refer to the procedure given for alternator drivebelt checking and adjustment under the *'ELECTRICAL SYSTEM'* heading on page 102.

IGNITION SYSTEM

Renew the spark plugs

Note: *Ensure that the correct type of new spark plugs is obtained for your particular car – refer to* ***'Service specifications'*** *on page 67. A suitable spark plug spanner and a spark plug gap adjustment tool, or a set of feeler gauges will be required for this task.*

The spark plugs must be renewed with the engine **cold**.

On some models, it may be necessary to remove the air cleaner or the air inlet hose to improve access to the spark plugs – refer to the air cleaner filter element renewal procedure in the *'24 000 miles (40 000 km)'* service Section on page 105 for details of air cleaner removal. If necessary, the air inlet hose can be removed by loosening the screws on the hose clips.

Mark the HT leads to make sure that they're refitted in their correct positions, then pull the leads from the spark plugs. Alternatively, only remove one HT lead at a time, and refit it once you've changed each spark plug. Pull on the rubber insulators at the ends of the leads, not on the leads themselves.

Before removing the spark plugs, brush any grit from the area around the plug recesses, to avoid it dropping into the engine as the plugs are removed.

Using a spark plug spanner, unscrew the spark plugs and remove them from the engine.

Before fitting the new plugs, the electrode gap must be set as follows (refer to *'Service specifications'* on page 67 for the correct electrode gap).

Measure the gap between the electrodes using a spark plug gap adjustment tool, or a feeler gauge. The gap is correct when the tool or gauge is a firm sliding fit between the electrodes.

If adjustment is required, carefully bend the **outer** electrode until the correct gap is obtained. **Never** try to bend the centre electrode.

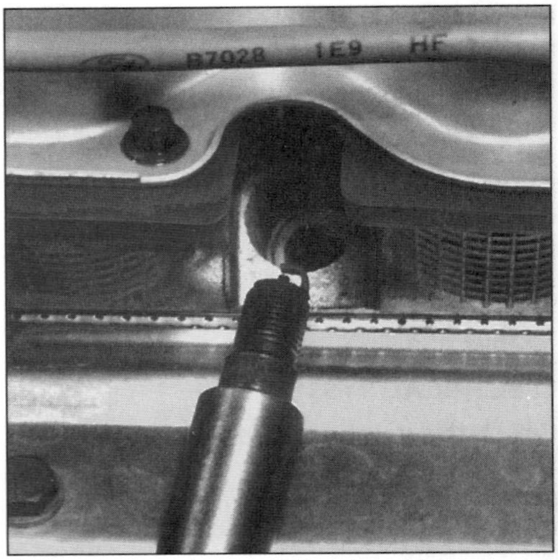

▲ *Removing a spark plug – CVH engine model*

▲ *Checking a spark plug gap using a suitable tool*

▲ *Adjusting a spark plug gap using the correct tool*

Make sure that the plug threads and the seating areas in the cylinder head are clean, and apply a little light oil to the plug threads. Screw the plugs in by hand initially.

Using the spark plug spanner, tighten each plug until initial resistance is felt. Tighten by a further one-sixteenth of a turn for the taper-seat type plugs fitted to OHC engines, and to V6 engines fitted with a catalytic converter. Tighten by a further one-quarter of a turn for the washer type plugs fitted to DOHC engines, CVH engines and non-catalyst V6 engines. **Do not** overtighten the plugs.

Refit the HT leads in their correct positions as noted before removal, making sure that they are a secure fit over the ends of the plugs.

Where applicable, refit the air cleaner or the air inlet hose using a reversal of the removal procedure.

AUTOMATIC TRANSMISSION

Check and if necessary top-up the automatic transmission fluid level

To check the fluid level, the car should be standing on level ground.

Open the bonnet and look for the fluid level dipstick, which is located on the right-hand side of the transmission to the rear of the engine (when viewed from the driver's seat).

Make sure that the handbrake is applied, then start the engine and allow it to idle. Apply the brake pedal, then move the transmission selector lever through all its positions three times, finishing in position **P**.

Wait for approximately one minute, then with the engine still idling, withdraw the fluid level dipstick, wipe it clean (using a clean, non-fluffy cloth), then slowly push it back into its tube, and pull it out again.

The fluid level should be between the **MIN** and **MAX** marks.

If necessary, top-up the level through the dipstick tube using the correct type of fluid (refer to *'Service specifications'* on page 68), then recheck the level.

Refit the dipstick and switch off the engine on completion.

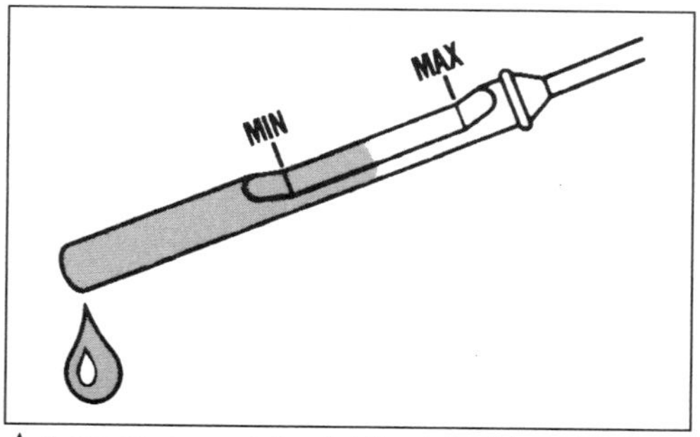

▲ *Automatic transmission fluid level dipstick markings*

BRAKING SYSTEM

Check the operation of the handbrake mechanism

Engage 1st gear (manual gearbox models) or **P** (automatic transmission models), as applicable, then chock the front wheels.

Jack up the rear of the car, and support it securely on axle stands (refer to *'Safety first!'* on page 88 for details of the correct jacking procedure).

Apply the handbrake and check that both rear wheels are locked, then release the handbrake and check that both rear wheels can be turned by hand (there may be some resistance, but it should be possible to turn the wheels without excessive force).

If the wheels do not lock with the handbrake applied, or cannot be turned with the handbrake released, the handbrake mechanism is faulty (possibly due to lack of lubrication or dirty components), and advice should be sought.

Check that all the components move freely and operate correctly, and wipe away any excess grease.

On completion, lower the car to the ground and remove the chocks from the front wheels.

BODYWORK

Lubricate all hinges, door locks and the bonnet release mechanism

Use general-purpose light oil for the door locks, and a suitable general-purpose grease for the hinges and bonnet release mechanism.

Don't use an excess of lubricant, as it may find its way onto other components, or onto the clothing of the car's occupants!

Check the operation of all door, tailgate/boot, bonnet release and window regulator components

If any of the components prove to be faulty, seek advice. Have the problem fixed as soon as possible, and without delay if the problem is likely to affect the car's security and safety (a faulty lock, or a window which can't be closed etc).

ELECTRICAL SYSTEM

Check the condition and adjustment of the alternator drivebelt

Note: *If the drivebelt requires adjustment, a suitable spanner or socket will be required to fit the alternator mounting and adjuster link bolts*

Examine the drivebelt for signs of cracking, obvious wear, or contamination, and renew it if necessary (note that on all except CVH engines, the belt also drives the engine coolant pump, and on certain models it may also drive the power steering pump).

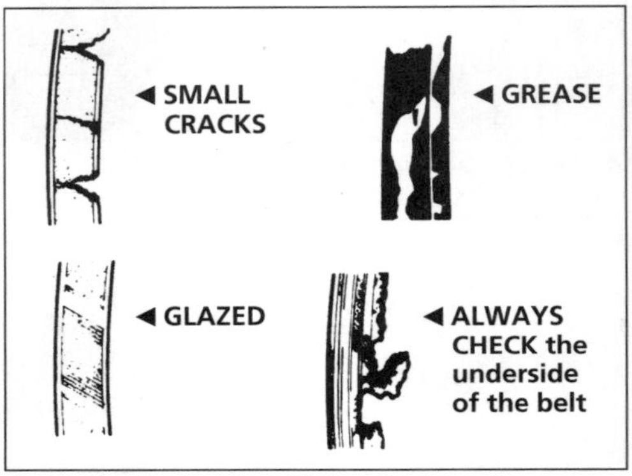

▲ *Examine the alternator drivebelt for signs of wear, deterioration or contamination – V-belt*

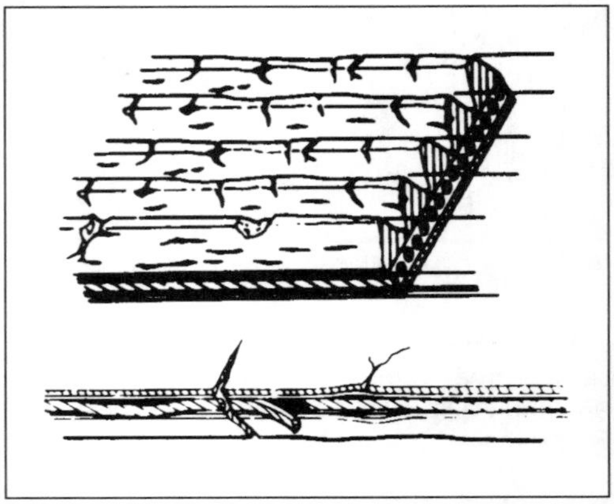

▲ *Examine the alternator drivebelt for signs of wear, deterioration or contamination – multi-ribbed belt*

Note that on some DOHC engine models, an automatic belt tensioner is fitted, and there is no need to check the belt tension on these models. The only DOHC engine models on which the belt tension should be checked are models with three belt pulleys (coolant pump, alternator and crankshaft); if four or more pulleys are fitted, there is no need to check the tension.

To check the tension of the belt, press down on the belt at the midway point of its longest run (between the coolant pump and alternator or power steering pump pulleys, depending on model).

▲ *Alternator drivebelt tension checking point (arrowed) – OHC and DOHC engine models without power steering*

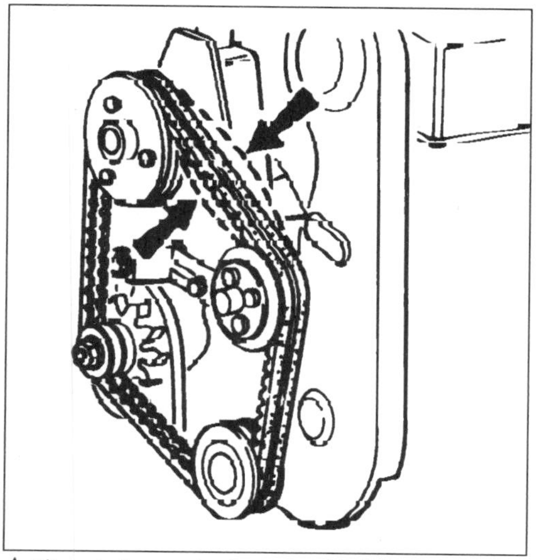

▲ *Alternator/power steering pump drivebelt tension checking point – (arrowed) OHC engine model*

The deflection of the belt should be approximately 10.0 mm (0.4 in) on all except 1.6 litre CVH and V6 engine models, and 4.0 mm (0.16 in) on 1.6 litre CVH and V6 engine models, under moderate pressure.

To adjust the tension, slacken the alternator mounting and adjuster link nuts/bolts, and move the alternator towards or away from the engine as necessary to give the correct tension. Note that the alternator mounting arrangement differs depending on model, but on OHC engine models, the mounting and adjuster link nuts/bolts should be tightened in the order shown in the accompanying illustration.

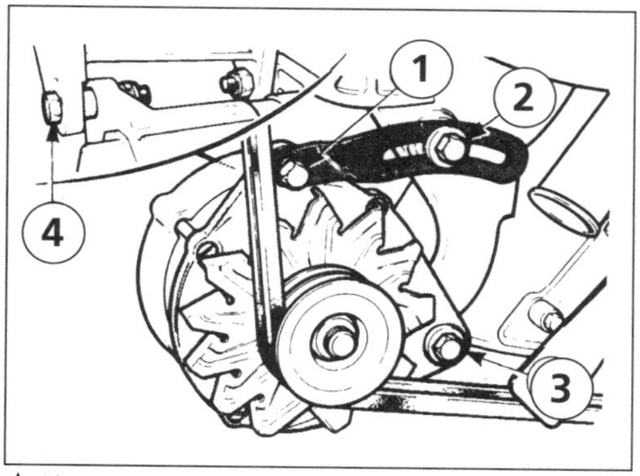

▲ *Alternator mounting and adjuster link nuts/bolts tightening sequence – OHC engine model (twin belt layout shown, as fitted to models with power steering)*

Once the tension is correct, tighten the adjuster link and alternator mounting nuts/bolts.

For details of the drivebelt renewal procedure, refer to the Owners Workshop Manual.

ROAD TEST

Check the operation of all instruments and electrical equipment

Make sure that all instruments read correctly, and switch on all electrical equipment in turn to check that it functions properly.

Check for any abnormalities in the steering, suspension, handling or road 'feel'

Drive the car and check that there are no unusual vibrations or noises.

Check that the steering feels positive, with no excessive 'sloppiness' or roughness, and check for any suspension noises when cornering and driving over bumps.

Check the performance of the engine, clutch (where applicable), gearbox and driveshafts

Listen for any unusual noises from the engine, clutch, gearbox and rear axle.

Make sure that the engine runs smoothly when idling, and that there's no hesitation when accelerating.

Check that, where applicable, the clutch action is smooth and progressive, that the drive is taken up smoothly, and that the pedal travel is not excessive. Also listen for any noises when the clutch pedal is depressed.

On manual gearbox models, check that all gears can be engaged smoothly without noise, and that the gear lever action is not abnormally vague or 'notchy'.

On automatic transmission models, make sure that all gearchanges occur smoothly without snatching, and without an increase in engine speed between changes. Check that all the gear positions can be selected with the car at rest. If any problems are found, they should be referred to a Ford dealer.

On four-wheel-drive models, listen for a metallic clicking sound from the front of the car as the car is driven slowly in a circle with the steering on full lock. Carry out this check in both directions. If a clicking noise is heard, this indicates wear in a front driveshaft joint, in which case seek advice and renew the joint if necessary.

Check the operation and performance of the braking system

Make sure that the car does not pull to one side when braking, and that the wheels do not lock prematurely when braking hard.

Check that there's no vibration through the steering when braking.

Check that the handbrake operates correctly, without excessive movement of the lever, and that it holds the car stationary on a slope.

ADDITIONAL TASKS

Note that as well as the tasks described in the preceding paragraphs, the additional tasks given in the *'Service schedule'* chart on page 82 should be carried out every 12 000 miles (20 000 km) or 12 months – whichever comes first. These tasks require more detailed explanation, or the use of special tools, and are considered beyond the scope of this Handbook. For details of these additional tasks, refer to the Owners Workshop Manual.

Every 24 000 miles (40 000 km) or 2 years – whichever comes first

In addition to the items in the *'12 000 miles (20 000 km)'* service on page 100 and *'6000 miles (10 000 km)'* service on page 95, carry out the following.

ENGINE

Renew the crankcase emission valve, where applicable (OHC and V6 engines only)

On OHC engines, the crankcase emission valve (where fitted) is located in the plastic oil

▲ *Removing the crankcase emission valve from the oil separator tube – OHC engine*

separator tube on the left-hand side of the engine (when viewed from the driver's seat), just in front of the oil level dipstick.

On V6 engines, the valve (where fitted) is located at the rear of the right-hand engine rocker cover (when viewed from the driver's seat).

▲ Removing the crankcase emission valve from the rocker cover – V6 engine

Pull the valve from its location in the oil separator tube or the rocker cover, as applicable, then release the clip and pull the valve from the end of the hose.

Fit the new valve using a reversal of the removal procedure. Make sure that the hose is securely connected, and that the valve is a secure fit in the oil separator or rocker cover, as applicable.

FUEL SYSTEM

Renew the air cleaner filter element

Carburettor engine models and 1.6 litre CVH engine models

The air cleaner filter element is located in the circular air cleaner casing on top of the carburettor/throttle housing.

Remove the securing screws from the top of the air cleaner cover.

Where applicable, release the spring clips around the edge of the cover, then pull off the cover.

Lift out the element.

▲ *Loosening an air cleaner cover securing screw – carburettor engine model*

▲ *Release the spring clips (where applicable) ...*

▲ *... and lift off the air cleaner cover ...*

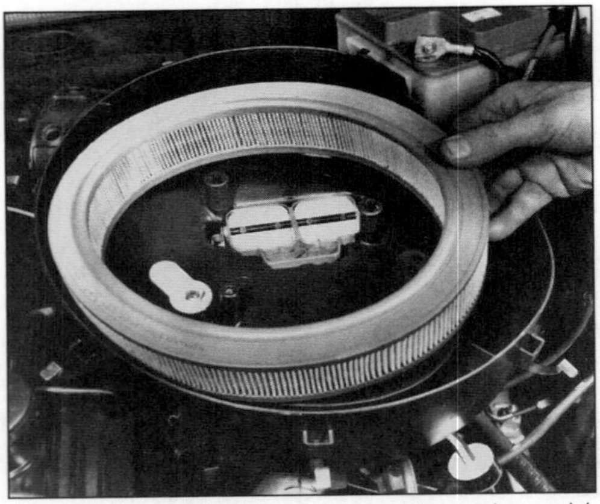

▲ ... for access to the element – carburettor engine model

Wipe clean the inside of the air cleaner body, taking care not to allow dirt to drop into the carburettor/throttle housing. Also clean the inside of the cover.

If desired, the air cleaner body can be removed for cleaning after disconnecting the various hoses and wiring plugs, as applicable. Make sure that the ignition is switched off before disconnecting any wiring, and note that on some OHC engine models, the air cleaner body is secured to the top of the engine with an additional screw.

Place the new element in position.

Refit the air cleaner (where applicable) and the cover by reversing the removal operations.

▲ Disconnecting the airflow meter wiring plug – OHC fuel injection engine model

OHC fuel injection engine models

The air cleaner is located at the front left-hand side of the engine compartment (when viewed from the driver's seat).

Disconnect the battery negative (black) lead, then press the locking clip on the airflow meter wiring plug, and disconnect the plug. Pull on the plug, not on the wiring.

Loosen the securing clip, and disconnect the air intake hose from the airflow meter on the side of the air cleaner cover.

Release the securing clips from the edge of the air cleaner cover, then lift off the cover complete with the airflow meter.

▲ Air cleaner cover securing clip (arrowed) – OHC fuel injection engine model

▲ Lifting out the air cleaner element – OHC fuel injection engine model

Lift out the element, then wipe clean the inside of the air cleaner casing and the cover.
Fit the new element with the sealing lip uppermost.
Refit and reconnect the remaining components, using a reversal of the removal procedure, but make sure that the air intake hose clip is correctly aligned as shown in the accompanying illustration.

DOHC and 2.8 litre V6 engine models

The air cleaner is located at the front left-hand side of the engine compartment (when viewed from the driver's seat).
Loosen the securing clip, and disconnect the air intake hose from the air cleaner cover.
Release the securing clips from the edge of the air cleaner cover, then lift off the cover.

▲ *Air cleaner cover securing clip (arrowed) – 2.8 litre V6 engine model*

Lift out the element, then wipe clean the inside of the air cleaner casing and the cover.
Place the new element in position.
Refit the cover using a reversal of the removal procedure.

2.9 litre V6 engine models

The air cleaner is located at the front left-hand side of the engine compartment (when viewed from the driver's seat).
Disconnect the battery negative (black) lead, then disconnect the wiring plugs from the two airflow meters on the side of the air cleaner cover.
Loosen the securing clips, and disconnect the air intake hoses from the airflow meters.
Release the securing clips from the edge of the air cleaner cover, then lift off the cover complete with the airflow meters.

▲ *Lifting off the air cleaner cover/airflow meters assembly – 2.9 litre V6 engine model*

Lift out the element, then wipe clean the inside of the air cleaner casing and the cover.
Place the new element in position.
Refit the cover using a reversal of the removal procedure.

On carburettor models, check the operation of the air cleaner air intake temperature control

All except CVH engine models

To check the operation of the air intake temperature control, the engine must initially be cold.
Loosen the securing clip, and disconnect the cold air intake hose from the air cleaner spout.
Look into the air cleaner spout. It should be possible to see all the way along the spout into the main air cleaner body (a small vertical rod which operates the temperature control flap should be visible across the inside of the spout).
Start the engine and allow it to idle, then look into the end of the air cleaner spout again (take care not to touch any of the hot engine components). The flap should now have moved across the spout to block off the supply of cold air.

If the flap does not move as described, the temperature control unit is faulty, and advice should be sought.

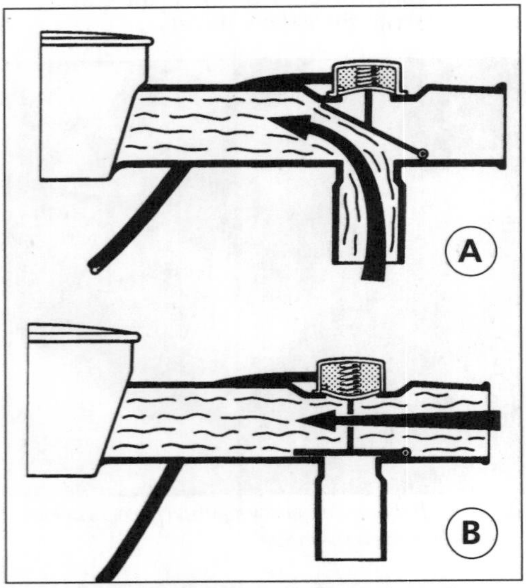

▲ *Air cleaner air intake temperature control operation – all except CVH engine models*
***A** Flap fully open to admit hot air*
***B** Flap fully closed to admit cold air*

CVH engine models

To check the operation of the air intake temperature control, the engine must initially be cold.

Loosen the securing clip, and disconnect the hot air intake hose from the air cleaner spout. It should be possible to see a flap across the hot air intake which should be fully open to allow hot air to enter the air cleaner.

Reconnect the hose, then start the engine and allow it to reach normal operating temperature.

Disconnect the hot air intake hose again, and look at the position of the flap, which should be fully closed to shut off the hot air intake.

If the flap positions are not as described, the temperature control is faulty, and advice should be sought.

IGNITION SYSTEM

Clean the distributor cap (where applicable), coil and HT leads

Note: *1.6 litre CVH engine models are fitted with a distributorless ignition system – on these models, the spark plug HT leads are connected directly to the ignition coil, and a distributor is not used.*

Where applicable, remove the distributor screening shroud from the top of the distributor, and disconnect the earth strap to

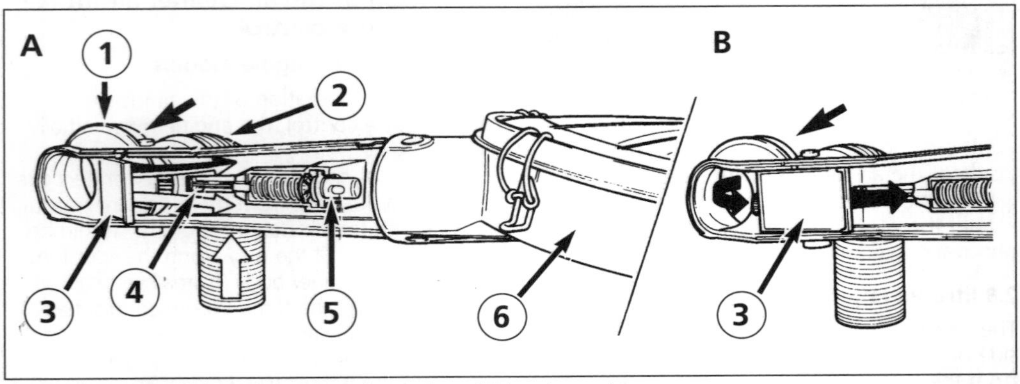

▲ *Air cleaner air intake temperature control operation – CVH engine models*
***A** Flap fully open to admit hot air*
***1** Air intake spout*
***2** Hot air intake hose*
***3** Flap valve*
***B** Flap fully closed to admit cold air*
***4** Link arm*
***5** Waxstat (temperature-sensitive operating mechanism)*
***6** Air cleaner body*

enable access to the distributor cap (some distributors may not be fitted with screening shrouds).

On OHC and 2.8 litre V6 engine models, the screening shroud can simply be unclipped from the distributor.

▲ *Unclipping the distributor screening shroud – OHC engine model*

On DOHC engine models, the screening shroud is secured by clips and screws, as shown in the accompanying illustration.

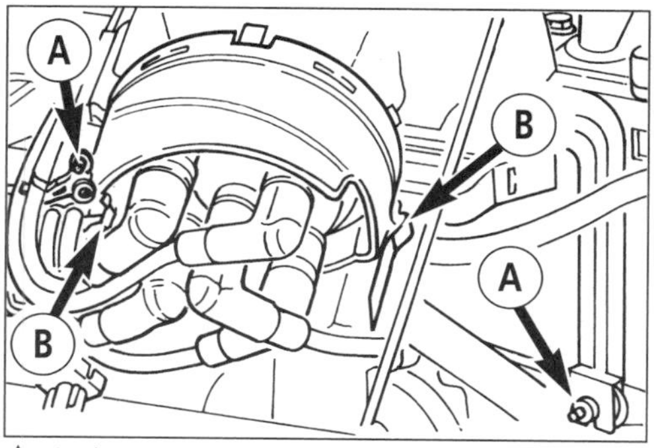

▲ *Distributor screening shroud – DOHC engine model*
A *Shroud securing screws*
B *Shroud securing clips*

On 1.8 litre CVH engine models, the plastic shroud can be removed after pulling the two halves of the shroud apart, and disconnecting the earth lead from the front of the engine.

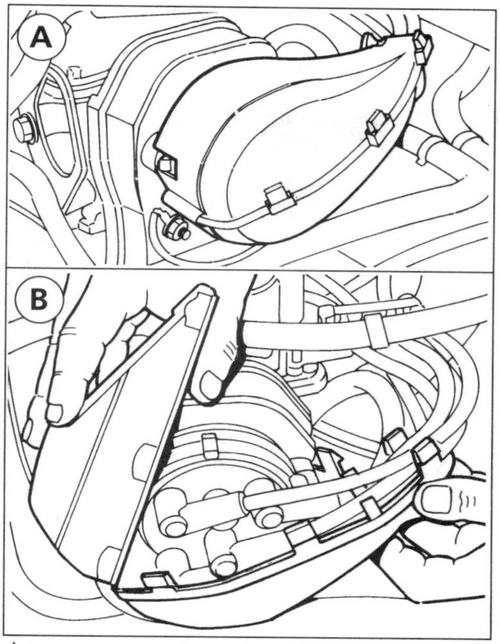

▲ *Removing the distributor shroud – 1.8 litre CVH engine model*
A *Shroud in position*
B *Separating the two halves of the shroud*

On 2.9 litre V6 engine models, there is no need to remove the distributor screening shroud, as it can be removed complete with the distributor cap.

▲ *Releasing a distributor cap securing clip – OHC carburettor engine model*

Remove the distributor cap, which may be secured with spring clips or screws, depending on model.

▲ *Removing a distributor cap securing screw – OHC fuel injection engine model*

▲ *Removing the distributor cap complete with screening shroud – 2.9 litre V6 engine model*

All except 1.6 litre CVH engine models

Thoroughly clean the distributor cap inside and out with a dry lint-free cloth.

Examine the inside of the cap, and if any of the four HT lead contacts inside the cap appear badly burnt or pitted, or if there are any signs of hairline cracks in the plastic, renew the cap.

Make sure that the carbon contact in the centre of the cap is not excessively worn. The contact should be free to move, and should protrude from its holder. If the contact is worn or damaged, renew the cap.

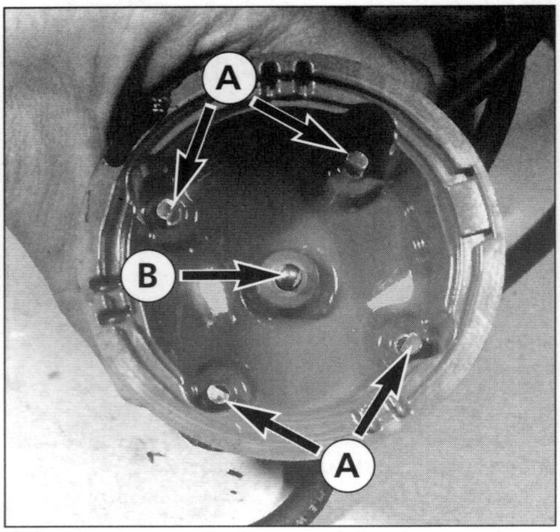

▲ *Distributor cap HT lead contacts (**A**) and carbon contact (**B**)*

Identify the spark plug leads by wrapping sticky labels around them, numbered from (for example) 1 to 4, counting from the front of the engine. Trace each plug lead through to the connection on the distributor cap, and attach a numbered sticky label next to the connection on the cap. Similarly identify the coil HT lead, if wished. On carburettor models, it may be necessary to remove the air cleaner from the top of the carburettor for access to the spark plugs – refer to the air cleaner filter element renewal procedure earlier in this Section for details.

Disconnect the HT leads from the coil, distributor cap, and the spark plugs. Wipe the coil clean, and check the contacts on the coil, distributor cap and the ends of the HT leads for corrosion.

Carefully clean away any corrosion, and wipe the leads clean over their entire length before refitting.

Make sure that the HT leads are refitted in their correct positions, as noted before removal.

1.6 litre CVH engine models

These models are fitted with a distributorless ignition system, so all that is required is that the spark plug HT leads and their connections to the ignition coil are kept clean.

It may be necessary first to remove the air cleaner for access to the spark plug leads – refer to the air cleaner filter element renewal procedure earlier in this Section for details.

It's probably easiest in this case to remove and clean one HT lead at a time – this will minimise confusion when reconnecting. Remove one HT lead from its spark plug and trace the lead through to the connection on the ignition coil. For access to the coil HT connections, remove the two screws and lift off the coil plastic shroud. Squeeze together the two legs on the ignition coil connector to remove the HT lead. Clean away any corrosion and dirt on and around the lead and coil connections, and wipe the lead over its entire length before refitting it. Repeat these operations for the remaining three HT leads in turn. Finally, wipe the coil body and the plastic shroud clean, and refit the shroud to complete.

STEERING

Check the condition and adjustment of the power steering pump drivebelt (where applicable)

The power steering pump may be driven by the same belt which drives the alternator, or it may be driven by a separate belt, depending on model.

If the pump is driven by the alternator drivebelt, refer to the procedure given for alternator drivebelt checking and adjustment under *'ELECTRICAL SYSTEM'* in the *'12 000 mile (20 000 km)'* service Section on page 102.

If the pump is driven by a separate drivebelt, proceed as follows:

Examine the drivebelt for signs of cracking, obvious wear, or contamination, and renew it if necessary.

To check the tension of the belt, press down on the belt at the midway point of its longest run.

The deflection of the belt should be approximately 4.0 mm (0.16 in), under moderate pressure.

To adjust the tension, slacken the power steering pump bracket securing bolts, or the pump pivot and adjustment bolts, as applicable, and move the pump as necessary to give the correct tension.

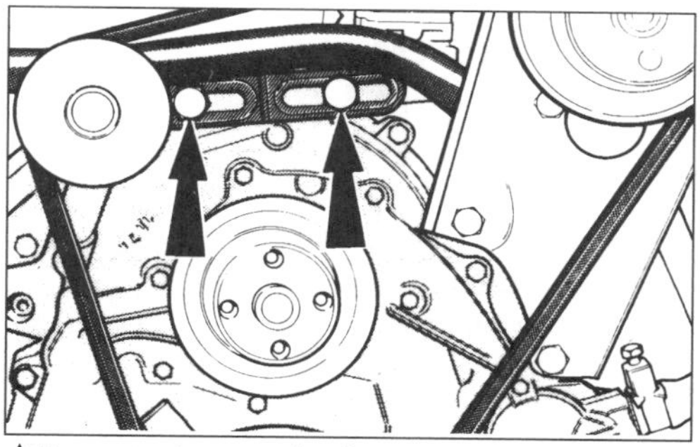

▲ *Power steering pump bracket securing bolts (arrowed) – certain V6 engine models*

Once the tension is correct, tighten the bracket securing bolts, or the pump pivot and adjustment bolts as applicable.

For details of the drivebelt renewal procedure, refer to the Owners Workshop Manual.

ADDITIONAL TASKS

Note that as well as the tasks described in the preceding paragraphs, the additional tasks given in the *'Service schedule'* on page 82 chart should be carried out every 24 000 miles (40 000 km) or 2 years – whichever comes first. These tasks require more detailed explanation, or the use of special tools, and are considered beyond the scope of this Handbook. For details of these additional tasks, refer to the Owners Workshop Manual.

Every 36 000 miles (60 000 km)

In addition to the items listed in the previous services, the additional engine task given at the end of the *'Service schedule'* chart on page 82, should be carried out every 36 000 miles (60 000 km). For details, refer to the Owners Workshop Manual.

Every 36 000 miles (60 000 km) or 2 years – whichever comes first

COOLING SYSTEM

Renew the coolant

Note: *The following items will be required for this task:*

- *A suitable quantity of coolant solution (made up from approximately 55% clean water and 45% antifreeze)*
- *A suitable container to catch the old coolant as it drains*
- *A spanner or socket to fit the cylinder block drain plug (where applicable)*

Draining the system

The coolant should be drained with the engine cold.

Remove the cooling system pressure cap from the expansion tank. Loosen the cap slowly in stages to gradually release any pressure in the system.

Certain OHC engine models have a bleed spigot on the thermostat housing (located at the front right-hand side of the engine when viewed from the driver's seat), which is covered by a rubber cap. The cap should be removed from the spigot before draining the system.

▲ *Loosening the rubber cap securing clip on the thermostat housing bleed spigot – OHC engine model*

Check to see if a drain plug is fitted to the lower right-hand or left-hand side of the radiator.

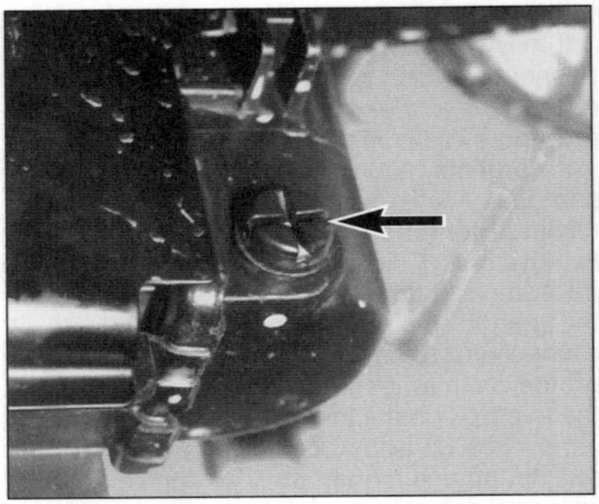

▲ *Radiator drain plug (arrowed) – CVH engine model*

If a drain plug is fitted, position a suitable container underneath the radiator, making sure that the container is big enough to hold all the coolant, then unscrew the drain plug and allow the coolant to drain.

If no drain plug is fitted, position the container beneath the radiator bottom hose connection, then slacken the hose clip, release the hose, and allow the coolant to drain.

▲ *Slackening the radiator bottom hose clip – OHC model*

A cylinder block drain plug is fitted to certain models as follows:

OHC engines – lower rear right-hand side (when viewed from the driver's seat) of the engine.

DOHC engines – no drain plug.

CVH engines – no drain plug.

2.3 and 2.8 litre V6 engines – towards the lower rear right-hand side of the engine (when viewed from the drivers seat), just in front of the oil filter.

2.9 litre V6 engines – towards the lower rear left-hand side (when viewed from the driver's seat) of the engine, above the starter motor (cover the starter motor with a plastic bag before removing the drain plug).

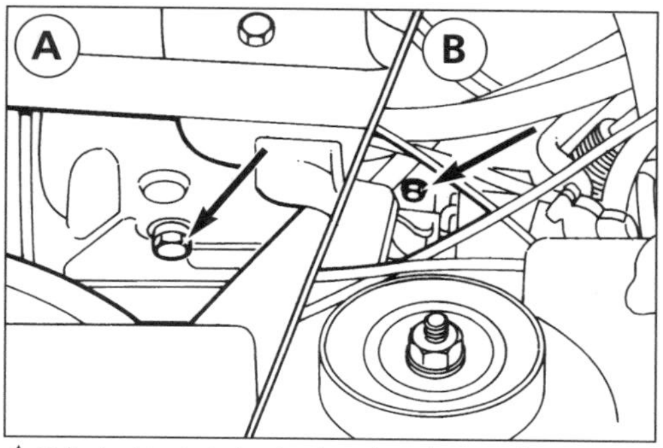

▲ *Cylinder block drain plug locations (arrowed)*
A *OHC engine models*
B *2.3 and 2.8 litre V6 engine models*

Where a drain plug is fitted, reposition the container, then unscrew the plug and allow the remaining coolant in the cylinder block to drain into the container.

If rust or sludge is evident in the coolant which has been drained, the cooling system should be flushed to remove any further contamination from the engine and radiator.

If the coolant drained is clean, then the cooling system can be refilled as described later in this Section.

Flushing the system

To flush the radiator, disconnect the top hose from the radiator, then insert a garden hose into the top of the radiator. Allow water to circulate until it runs clear from the bottom of the radiator.

On all except V6 engine models, to flush the engine, insert the garden hose into the expansion tank filler neck. Allow water to circulate until it runs clear from the disconnected radiator bottom hose or radiator drain plug, and the cylinder block drain plug as applicable.

On V6 engine models, to flush the engine, insert the garden hose into the disconnected radiator top hose. Allow water to circulate until it runs clear from the disconnected radiator bottom hose or radiator drain plug, and the cylinder block drain plug as applicable.

Reconnect the top hose to the radiator and tighten the securing clip on completion.

Filling the system

Reconnect the radiator bottom hose and tighten the clip, or refit and tighten the radiator drain plug, and/or the cylinder block drain plug, as applicable.

On OHC models fitted with a bleed spigot on the thermostat housing, make sure that the rubber cap is removed.

A solution of 55% clean water and 45% antifreeze should be used to fill the system all year round. It's important to note that because of the different types of metals used in the engine, it's vital to use antifreeze with suitable anti-corrosion additives all year round. **Never** use plain water to fill the whole system. If there's any doubt about the strength of the coolant mixture, it's better to add too much antifreeze than not enough.

Pour coolant in through the expansion tank until the level is up to the **MAX** mark.

Where applicable on OHC models, refit the rubber cap to the bleed spigot on the thermostat housing when coolant starts to flow from the spigot. Make sure that the clip is tight.

Wait a few moments for trapped air to escape, then add more coolant to the expansion tank.

Squeeze the coolant hoses to help to disperse air locks, then top-up the coolant level if necessary. Repeat this procedure until the level does not drop, then refit the pressure cap.

Start the engine and run it to normal operating temperature, then switch off.

Once the engine has cooled, squeeze the hoses and check for air locks (it should be possible to feel coolant in all the hoses).

If necessary, air can be removed from a hose by loosening a hose clip, and squeezing the hose until all the air is forced out (obviously some coolant will be forced out too).

Recheck the coolant level, and carry out any final topping-up to the expansion tank.

Every 36 000 miles (60 000 km) or 3 years – whichever comes first

In addition to the items listed in the previous services, the additional braking system task given in the *'Service schedule'* chart on page 84 should be carried out every 36 000 miles (60 000 km) or 3 years – whichever comes first. For details, refer to the Owners Workshop Manual.

SEASONAL SERVICING

If you carry out all the procedures described in the previous Section, at the recommended mileage or time intervals, then you'll have gone a long way towards getting the best out of your car in terms of both performance and long life. In spite of this, there are always other areas, not dealt with in regular servicing, where neglect can spell trouble.

A little extra time spent on your car at the beginning and end of every winter will be well worthwhile in terms of peace of mind and prevention of trouble. The suggested tasks which follow have therefore been divided into Spring and Autumn Sections – but it's always a good idea to do them more frequently if you feel able.

Autumn

COOLING SYSTEM

Check the radiator and all hoses for signs of deterioration or damage

Refer to *'ENGINE COMPARTMENT'* in the *'6000 miles (10 000 km)'* service Section on page 95.

Check the strength of the coolant solution

This can be done using a coolant hydrometer, which should be available from most motor accessory shops and motor factors.

The hydrometer will give an indication of the amount of antifreeze in the system, and should be used exactly in accordance with the manufacturer's instructions.

The strength of the solution should never be allowed to drop below 55% clean water, and 45% antifreeze.

If there is any doubt about the strength of the coolant solution, the system should be drained and refilled as described in *'Service tasks'* on page 112.

ELECTRICAL SYSTEM

Where applicable, check the battery electrolyte level

Refer to *'Regular checks'* on page 78.

Check and if necessary clean the battery terminals

Make sure that the terminals are secure, and clean any corrosion from the metal.

To prevent corrosion, the terminals can be coated with petroleum jelly (don't use ordinary grease).

Check the condition and adjustment of the alternator drivebelt

Refer to the procedure given in the *'12 000 miles (20 000 km)'* service Section on page 102.

Check the operation of all lights, electrical equipment and accessories

Refer to *'Fault finding'* on page 131 if any problems are discovered.

If any of the bulbs need to be renewed, refer to *'Bulb, fuse and relay renewal'* on page 119.

Check the washer fluid level, and the wipers and washers

Refer to *'Regular checks'* on page 78.

TYRES

Check tread depth and condition

Refer to *'Regular checks'* on page 76. Remember that you may be driving in slippery conditions during the Winter.

BODYWORK

Thoroughly clean the car

Wash the car, and then polish it thoroughly to help protect the paint during the Winter.

Spring

UNDERBODY

Thoroughly clean the underside of the car

The best time to clean the underside of the car is after the car has been driven in wet conditions, when the accumulated dirt will be softened up.

To clean the car, first of all the car must be jacked up as high as possible (making sure that it is safely supported – refer to *'Safety first!'* on page 88).

Gather together a quantity of paraffin, or water-soluble solvent, a stiff-bristled brush, a scraper, and a garden hose.

With the car jacked up and safely supported, get underneath the car, and cover the brake components at each wheel with polythene bags to stop dirt and water getting into them.

Loosen any encrusted dirt, scraping or brushing it away – the paraffin or solvent can be used where there's oil contamination. Pay particular attention to the wheel arches. Take care not to remove the underseal (a waxy substance applied to the car underbody to protect against corrosion).

When all the dirt is loosened, a wash down with the hose will remove the remaining dirt and mud.

You can now check for signs of damage to the underseal. If there's any sign of the underseal breaking away, patch it up by spraying or brushing on a suitable underseal wax (available from motor accessory shops and motor factors). Make sure the area is clean and dry before applying the wax.

Take the opportunity to check for signs of rusting. Likely places are the body sills and floor panels.

If rust is found, seek advice and have the affected area repaired before the problem gets too bad.

On completion, lower the car to the ground.

BODYWORK

Thoroughly check and clean the surfaces of the bodywork

Give the car a thorough wash, and check for stone chips and rust spots. For details of how to treat these minor paint problems, refer to *'Bodywork and interior care'* on page 118.

After any repairs to the paint have been carried out (not before, otherwise the paint will not stick), give the car a polish.

TOOLS

WHAT TO BUY

If you're intending to carry out your own servicing, you'll need to obtain a few basic tools. Although at first sight you may think that tools seem expensive, once you've bought them they should last a lifetime if you look after them properly.

The tools supplied with the car will enable you to change a wheel, and not much more. The absolute minimum tool kit you'll need to carry out any maintenance or servicing will be a range of metric spanners, two screwdrivers (one for crosshead or 'Phillips' type screws), and a pair of pliers. With a bit of ingenuity, these items should enable you to complete the more basic routine servicing jobs, but they won't allow you to do much else.

When buying tools, it's important to bear in mind the quality. You don't need to buy the most expensive tools available, but generally you get what you pay for. Cheap tools may prove to be a false economy, as they're unlikely to last as long as better quality alternatives.

It's very difficult to lay down hard and fast rules on exactly what you're going to need, but the following list should be helpful in building up a good tool kit. Combination spanners (ring one end, open-ended the other) are recommended because, although more expensive than double open-ended ones, they give the advantages of both types.

- *Combination spanners to cover a reasonable range of sizes (say 7 mm to 17 mm at least)*
- *Adjustable spanner*
- *Spark plug spanner (with rubber insert)*
- *Spark plug gap adjustment tool*
- *Set of feeler gauges*
- *Screwdriver (Plain) – 100 mm long blade x 6 mm diameter (approx)*
- *Screwdriver (Crosshead) – 100 mm long blade x 6 mm diameter (approx)*
- *Oil filter wrench*
- *Pliers*
- *Tyre pump*
- *Tyre pressure gauge*
- *A suitable key (Torx or Allen type, depending on model) to fit the gear-box oil filler plug (later models only)*
- *Oil can*
- *Funnel (medium size)*
- *Stiff brush (for general cleaning jobs)*
- *Tool box*
- *Hydraulic jack (ideally a trolley jack)*
- *Pair of axle stands*
- *Suitable containers for draining engine oil and coolant*
- *Inspection lamp*

This is by no means a comprehensive list of tools, and you'll probably want to gradually add to your tool box as you discover the need for other tools, especially if you decide to tackle some of the more advanced servicing jobs described in the **Owners Workshop Manual** for your particular car.

CARE OF YOUR TOOLS

Having bought a reasonable set of tools, it's worth taking the trouble to look after them. After use, always wipe off any dirt or grease using a clean, dry cloth before putting them away. Never leave tools lying around after they've been used – a tool rack, or better still a proper toolbox will prove the best way of keeping everything up together. Rags can be wrapped around loose tools to prevent them from rattling if you're going to keep them in the boot of the car.

Feeler gauges should be wiped with an oily cloth from time to time, to leave a thin coating of oil on the metal which will prevent corrosion. Screwdriver blades inevitably lose their keen edges, and a little occasional attention with a file or an oilstone will keep them in good condition.

It's well worth spending the time and effort necessary to look after your car's bodywork and interior. Cleaning your car regularly will not only improve its appearance, it will also protect the bodywork against the elements and the grime encountered in everyday driving. It's worth bearing in mind that if your car looks clean and tidy, it will also be worth more money when you come to sell it or trade it in for a newer model.

This Section will help you to keep your car in 'showroom' condition. If you wish to tackle repairs to more serious bodywork damage or corrosion, comprehensive details can be found in our **'Car Bodywork Repair Manual'**.

CLEANING THE INTERIOR

It's a good idea to clean the interior of the car first, before cleaning the exterior, as this will avoid spreading the dirt from inside over the bodywork.

Start by removing all the loose odds and ends from inside the car, not forgetting the ashtrays, glovebox and any interior pockets. Take out any loose mats or carpets, which should be shaken and brushed, and if possible vacuum-cleaned.

The inside of the car can be cleaned with a brush and dustpan, or preferably a vacuum-cleaner. If you can't use your household vacuum-cleaner, it may be worth investing in one of the small 12-volt hand vacuum-cleaners which can be powered from the car battery. If the carpets are very dirty, use a suitable proprietary cleaner with a brush to remove the ingrained dirt. Ideally, it's best to remove the carpets for cleaning, but this is an involved task in most modern cars, and it will probably be easier to leave them in place.

The facia, door trim, seats, and any other items which require attention can now be wiped over with a cloth soaked in warm water containing a little washing up liquid. If the trim is particularly dirty, use one of the proprietary cleaners available from motor accessory shops – a number of different types are available, suitable for vinyl, cloth or leather upholstery, etc, as required. An old nail brush or tooth brush will help to remove any ingrained dirt. On completion, wipe the surfaces dry using a lint-free cloth, and leave the windows open to speed up drying.

The inside of the windscreen and windows should be cleaned using a proprietary glass cleaner, or methylated spirits. Be careful about using certain household products such as washing-up liquid which may leave a smeary film. Finish off by wiping with a clean, dry paper tissue.

Don't forget to clean the boot.

Check for any nicks or tears in the interior trim, seats, headlining, etc. Various repair kits are available from motor accessory shops to suit most types of trim, but if the damage is very serious, the relevant trim panel will probably have to be renewed.

CLEANING THE BODYWORK

Ideally, the car should be washed every week, either by hand (preferably using a hosepipe), or by using a local car-wash. If you're washing the car by hand, use plenty of water to loosen the dirt and dust, and if possible use a suitable car shampoo or wax additive in the water. Any stubborn dirt such as road tar or bird droppings can be removed using methylated spirit or preferably one of the special proprietary cleaning solutions (in which case make sure that the product is suitable for use on car paint) – in either case, wash the affected area down with plenty of water after removing the dirt. **Never** just wipe over a dirty car, as this will scratch the paint very effectively.

Two or three times a year, a good silicone or wax polish can be used on the paintwork. It's important to wash the car and remove all stubborn dirt, tar, etc, before using polish, as the polish will effectively seal over the top of any dirt, making it extremely difficult to remove in the future. Good polishes actually form a protective coating over the paint finish, which should help to make future accumulated dirt easier to remove. Always follow the manufacturer's recommendations closely when using polish. Try to avoid getting polish on any unpainted plastic or rubber body parts such as bumpers and spoilers, as it tends to leave a stain when dry. Chrome parts are best cleaned with a special chrome cleaner, as ordinary metal polish will wear away the finish.

Plastic or rubber body parts such as bumpers and spoilers should be cleaned using a suitable proprietary cleaner. A number of different types are available, including colour restorers, and special solvents to remove any stray polish which may have crept onto the plastic when polishing the paintwork. Make absolutely sure that any solvents or cleaners used are suitable, as certain products may attack plastic and/or rubber.

If the paint is beginning to lose its gloss or colour, and ordinary polishing doesn't seem to solve the problem, it's worth considering the use of a polish with a mild 'cutting' action to remove what is in effect a surface layer of 'dead' paint. In this case, follow the manufacturer's instructions, and don't polish too vigorously, or you might remove more paint than you intended!

DEALING WITH SCRATCHES

With superficial scratches which don't penetrate down to the metal, repair can be very simple.

Very light scratches can be polished out by carefully using a suitable 'cutting' polish. Follow the manufacturer's instructions, and take care not to remove too much of the surrounding paint.

If the scratch cannot be polished out, touch-up paint will be required. Touch-up paint is usually available in the form of a touch-up stick from the car manufacturer's dealers, or in various forms from motor accessory shops. You will probably have to quote the year and model of the car in order to make sure that you obtain the correct matching colour, and in some cases you may have to quote a paint reference number which will usually appear on the car's Vehicle Identification Number (VIN) plate under the bonnet (refer to *'Servicing'* on page 89).

Lightly rub the area of the scratch with a very fine cutting paste to remove loose paint from the scratch and to clear the surrounding bodywork of polish.

Rinse the area with clean water.

Apply suitable touch-up paint to the scratch using a fine paint brush, and continue to apply fine layers of paint until the surface of the paint in the scratch is level with the surrounding paintwork.

Allow the new paint at least two weeks to harden, then blend it into the surrounding paintwork by rubbing the scratch area with a paintwork renovator or a very fine cutting paste.

If the paint finish requires a lacquer coat (in which case the lacquer will be supplied with the touch-up paint), it should now be applied to the newly painted area and allowed to dry in accordance with the manufacturer's instructions.

Finally, apply a suitable wax polish to the affected area for added protection.

BULBS

A defective exterior light can be not only dangerous, but also illegal. Carrying spare bulbs will enable you to renew blown ones as they occur. A failed interior light bulb may be just a nuisance, but a faulty exterior light could be a life or death matter.

Before assuming that a bulb has failed, check for corrosion on the bulbholder and wiring connections, particularly around the rear light assemblies.

- **Note that as a safety precaution, the battery earth (black) lead should always be disconnected before renewing a bulb.**
- **Remember to reconnect the lead on completion.**

Bulb ratings

Always make sure that when a bulb is renewed, a new bulb of the correct rating is used.

All the bulbs are of the 12-volt type, and their ratings are as follows:

Bulb	Rating [watts]
Headlight	60/55
Front sidelights	5
Front and rear direction indicator lights	21
Front driving light	55
Front foglight	55
Front direction indicator repeater light	5
Rear brake/tail light (Hatchback models up to 1987)	21/5
Rear fog/tail light (all models from 1987 to 1990 except Estate)	21/5
Rear fog/tail light (all models from 1990 except Estate)	21/4
Rear brake light (all Estate models and all other models from 1987)	21
Tail light (Estate models)	5
Rear foglight (Hatchback models up to 1987 and all Estate models)	21
Reversing light	21
Rear number plate light	5
Courtesy light	10
Map reading light	5
Luggage compartment light	5
Glovebox light	2.5

Exterior light bulbs

HEADLIGHT

Note that on some V6 engine models, it may improve access to the left-hand headlamp if the air cleaner cover is removed – refer to the procedure for air cleaner element renewal in the *'24 000 mile (40 000 km)'* service Section on page 107.

Open the bonnet, and remove the protective cover from the rear of the headlight by turning it anti-clockwise.

Pull the wiring plug from the rear of the headlight bulb.

Rotate the bulb securing clip anti-clockwise, or release the spring clip arms from the bulb housing, according to type, and withdraw the bulb.

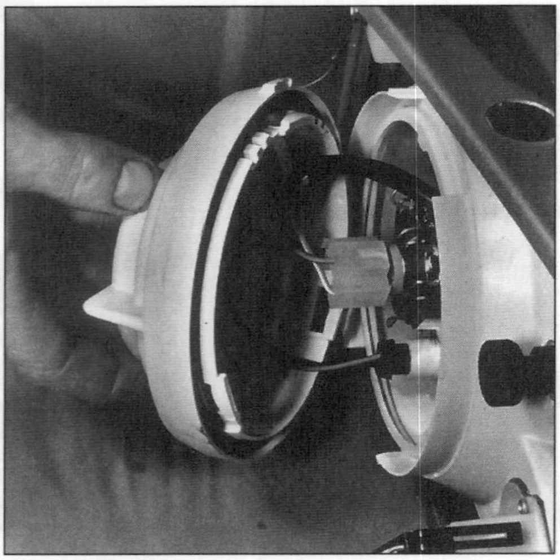

▲ *Removing the cover from the rear of the headlight*

▲ *Pulling the wiring plug from the headlight bulb*

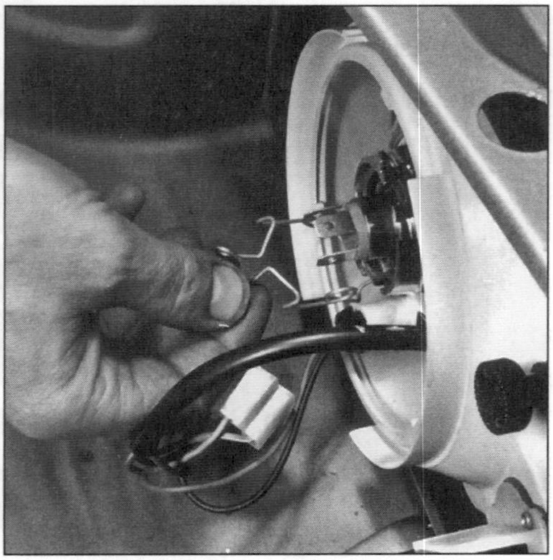

▲ *Releasing a headlight bulb securing spring clip*

▲ *Removing a headlight bulb*

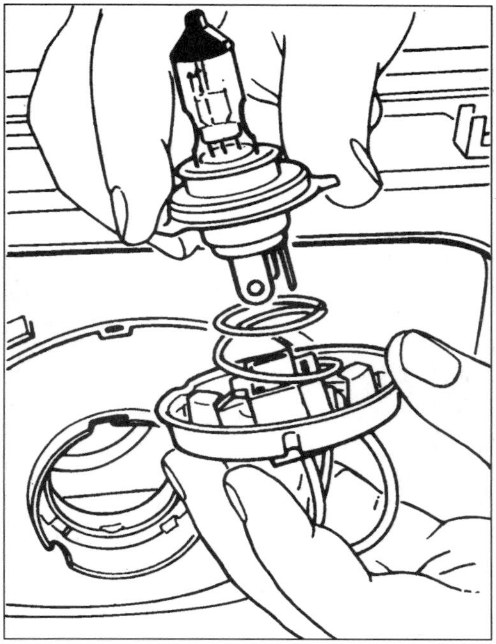

▲ *Removing a headlight bulb from its securing clip (alternative type)*

Fit the new bulb (which is an 'H4' type), taking care not to touch it with your fingers (hold the bulb using a clean cloth or paper tissue), using a reversal of the removal procedure. If the glass on the new bulb is inadvertently touched, wipe it with a cloth moistened with methylated spirit.

Make sure that the protective cover is securely fitted to the back of the headlight.

FRONT SIDELIGHT

Open the bonnet, and remove the protective cover from the rear of the headlight by turning it anti-clockwise.

Pull the sidelight bulbholder from the rear of the headlight, then pull the bulb from the holder.

Fit the new bulb using a reversal of the removal procedure. Make sure that the bulbholder is properly located in the rear of the headlight, and make sure that the protective cover is securely fitted to the back of the headlight.

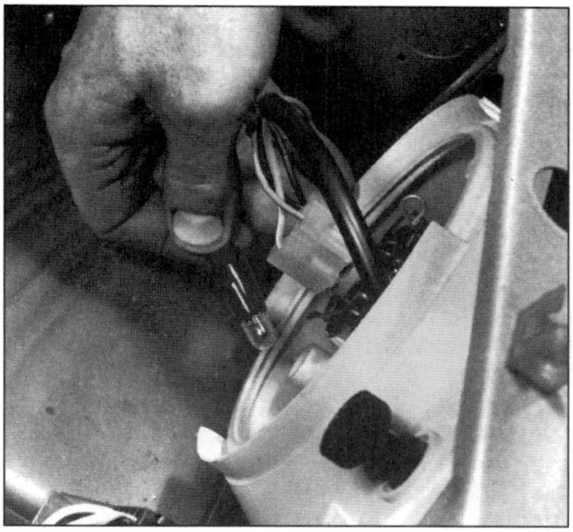

▲ *Removing a front sidelight bulbholder*

FRONT DIRECTION INDICATOR LIGHT Models up to 1987

On models not fitted with front foglights, gently push the indicator light unit rearwards into the bumper until the plastic retaining clip is heard to click into the locked position, when the light unit will spring forward out of the bumper opening.

On models with front foglights, push the release lever **(A)** upwards (the lever is hidden under the upper edge of the bumper), until the light unit springs forward out of the bumper opening.

▲ *Front direction indicator light release lever **(A)** – models without front foglights up to 1987*

Twist the bulbholder anti-clockwise to remove it from the light, then press in the bulb and turn it anti-clockwise to remove it from the holder.

Fit the new bulb using a reversal of the removal procedure. Make sure that the light unit retaining clip is in the released position, then push the light assembly evenly into the bumper until the retaining clip is heard to click into position.

Models from 1987

Open the bonnet, and unhook the direction indicator light unit retaining spring from the lug next to the headlight, then withdraw the light unit forwards and sideways from the car's front wing.

Twist the bulbholder anti-clockwise to remove it from the light, then press in the bulb and turn it anti-clockwise to remove it from the holder.

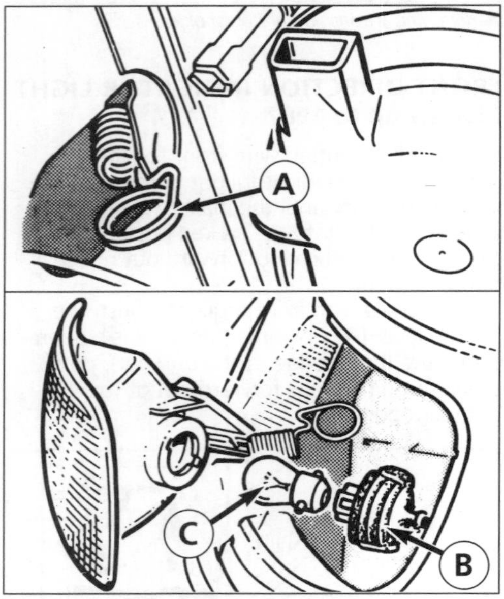

▲ *Front direction indicator light bulb renewal – models from 1987*
A *Light unit retaining spring*
B *Bulbholder*
C *Bulb*

Fit the new bulb using a reversal of the removal procedure. Make sure that the locating pins on the light unit engage in the holes in the front wing, and make sure that the light unit retaining spring is securely located on its lug.

FRONT DRIVING LIGHT

Open the bonnet, and remove the protective cover from the top of the driving light. On models up to 1987 the cover can be removed by twisting anti-clockwise, and on models from 1987 the cover can be removed after swinging the wire retaining clip forwards.

Release the bulb retaining spring clip by

▲ *Front driving light bulb cover (arrowed) – models up to 1987*

▲ *Removing the front driving light bulb – models up to 1987*

▲ *Front driving light bulb cover and retaining clip – models from 1987*

pushing it downwards or squeezing the legs together, depending on type, then withdraw the bulb (which is an 'H3' type with integral bulbholder and wire).

Pull the wiring out from the light, and disconnect at the connector plug.

Fit the new bulb, taking care not to touch it with your fingers (hold the bulb using a clean cloth or paper tissue), using a reversal of the removal procedure. If the glass on the new bulb is inadvertently touched, wipe it with a cloth moistened with methylated spirit.

Make sure that the protective cover is securely fitted to the top of the light.

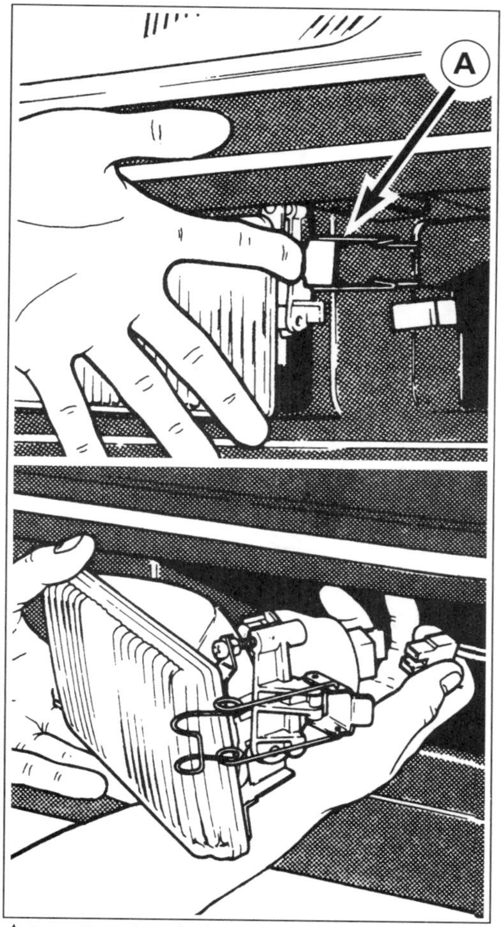

▲ *Removing a front foglight – models up to 1987*
A *Retaining tab*

FRONT FOGLIGHT

Models up to 1987

First, remove the front direction indicator light unit, as described in the Section for front direction indicator light bulb renewal for models up to 1987.

Release the foglight by pushing the retaining tab **(A)** rearwards, then withdraw the light, and disconnect the wiring plug.

Turn the bulb cover anti-clockwise to gain access to the bulb, then unhook the retaining clip, disconnect the wiring, and withdraw the bulb (which is an 'H3' type with integral bulbholder).

Fit the new bulb, taking care not to touch it with your fingers (hold the bulb using a clean cloth or paper tissue), using a reversal of the removal procedure. If the glass on the new bulb is inadvertently touched, wipe it with a cloth moistened with methylated spirit.

Make sure that the locating pins on the light unit engage in the bumper, and secure the unit in position with the retaining tab before refitting the direction indicator light unit.

Models from 1987

Unscrew the two retaining screws from the front of the light unit rim, and remove the lens complete with the reflector.

Pull the wire form the bulb terminal, then pull the spring clip from the rear of the bulb, and withdraw the bulb (which is an 'H3' type with integral bulbholder).

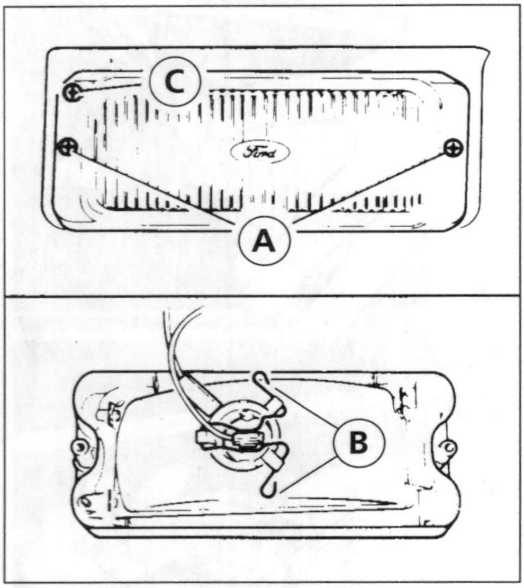

▲ *Front foglight bulb renewal – models from 1987*
A Lens securing screws
B Spring clip
C Foglight beam adjusting screw

Fit the new bulb, taking care not to touch it with your fingers (hold the bulb using a clean cloth or paper tissue), using a reversal of the removal procedure. If the glass on the new bulb is inadvertently touched, wipe it with a cloth moistened with methylated spirit.

Take care not to trap the wiring between the light unit and the bumper during reassembly.

FRONT DIRECTION INDICATOR REPEATER LIGHT

Turn the complete light assembly clockwise, and pull it from the front wing.

Detach the bulbholder from the light assembly by twisting it anti-clockwise and pulling it out, then pull the bulb from the holder.

Fit the new bulb using a reversal of the removal procedure.

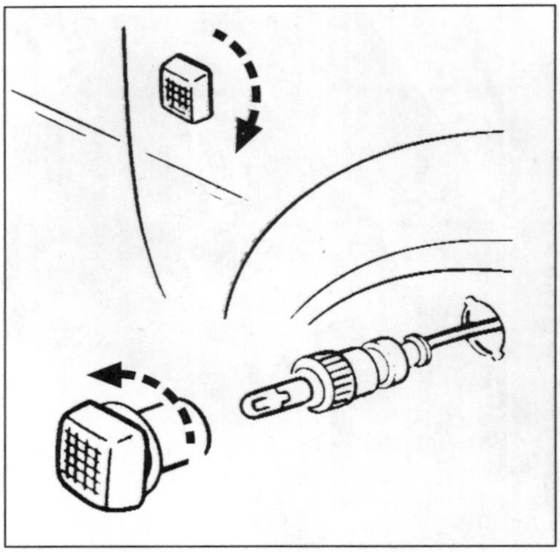

▲ *Front direction indicator repeater light bulb renewal*

REAR LIGHTS
Hatchback and Saloon models

Open the tailgate/boot lid, and press the retaining tab on the edge of the bulbholder assembly, then swing the assembly outwards to release the locating tab at the other end.

Remove the relevant bulb by pushing it down and turning it anti-clockwise.

Fit the new bulb and refit the bulbholder assembly, using a reversal of the removal procedure.

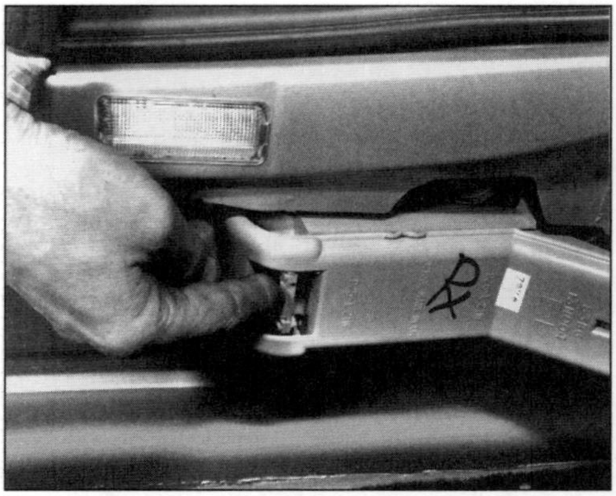

▲ *Press the rear light bulbholder retaining tab ...*

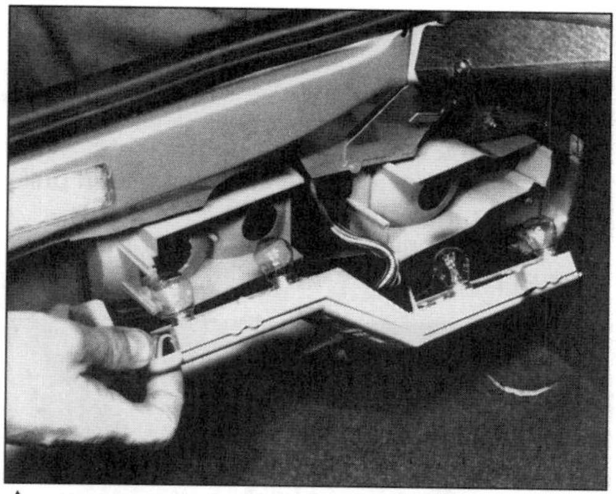

▲ ... to remove the bulbholder – Hatchback and Saloon models

Estate models

Open the tailgate, and where applicable remove the cover from the rear of the bulbholder assembly, either by turning the fasteners a quarter of a turn anti-clockwise using a coin, or by pulling the cover off, depending on type.

Push the top and bottom retaining tabs apart, and withdraw the bulbholder assembly.

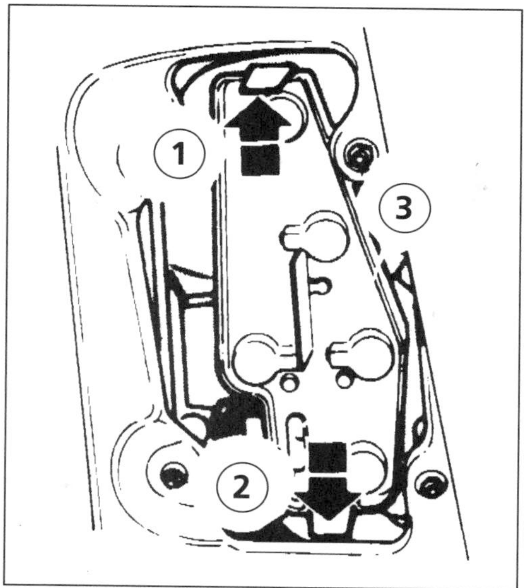

▲ *Push the retaining tabs* ***(1)*** *and* ***(2)*** *apart to remove the rear light bulbholder* ***(3)*** *– Estate models*

Remove the relevant bulb by pushing it down and turning it anti-clockwise.

Fit the new bulb and refit the bulbholder assembly, using a reversal of the removal procedure.

REAR NUMBER PLATE LIGHT

Using a small screwdriver, carefully prise the light assembly out through the top of the bumper. Take care to avoid scratching the bumper.

Turn the bulbholder anti-clockwise to remove it from the light assembly, then pull the bulb from the holder.

▲ *Twist the rear number plate light bulbholder* ***(B)*** *anti-clockwise to remove it from the light assembly* ***(A)***

Fit the new bulb, then reassemble and refit the light assembly by reversing the removal operations.

Interior light bulbs

Note: *Renewal of the various instrument panel and facia-mounted control illumination light bulbs requires detailed explanation, and in some cases extensive dismantling. Details can be found in the Owners Workshop Manual.*

COURTESY LIGHT

Using a small screwdriver, carefully prise the light assembly from its location in the roof, taking care not to damage the roof trim, then pull the bulb from the spring contacts in the light body. On models fitted with map reading lights, the reflector must be unclipped from the light for access to the bulb.

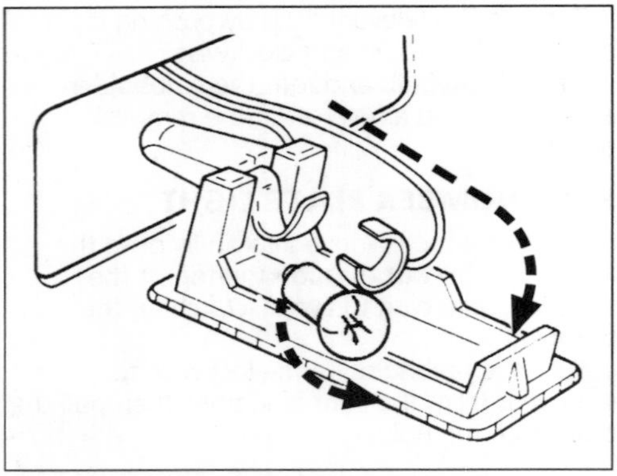

▲ *Renewing a courtesy light bulb – later models*

Fit the new bulb, making sure that it locates securely in the spring contacts, then refit the light assembly.

MAP READING LIGHT

Using a small screwdriver, carefully prise the courtesy light assembly from its location in the roof, taking care not to damage the roof trim, and disconnect the wiring.

Reach up into the courtesy light aperture, and push the map reading light down through the roof panel.

Pull the bulbholder from the map reading light, and then pull the bulb from the holder.

Fit the new bulb, then refit the light components using a reversal of the removal procedure.

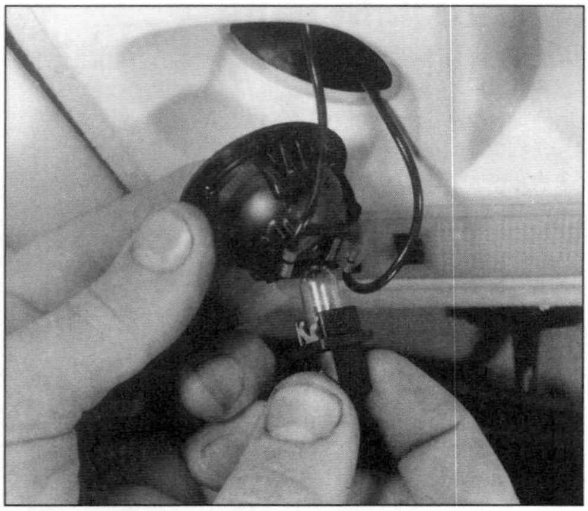

▲ *Removing a map reading light bulbholder from the light*

LUGGAGE COMPARTMENT LIGHT

Using a small screwdriver, carefully prise the light assembly from its location.

Remove the bulbholder from the light assembly by turning it anti-clockwise, then pull the bulb from the holder.

Fit the new bulb using a reversal of the removal procedure.

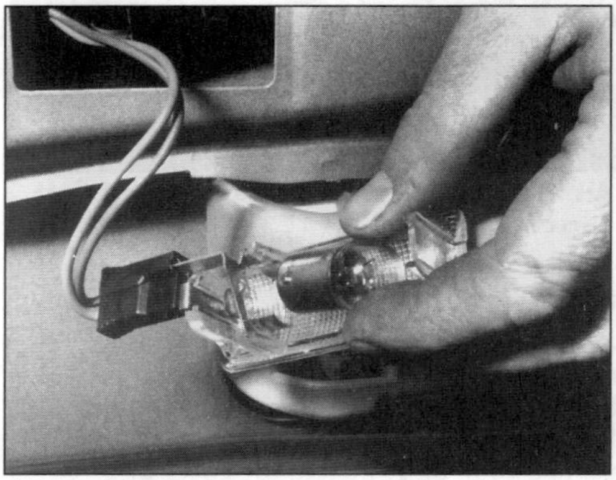

▲ *Renewing a luggage compartment light bulb*

GLOVEBOX LIGHT

Simply pull the bulb from its socket in the glovebox, and push the new one into place.

FUSES

Most of the car's electrical circuits are protected by fuses which will blow when the relevant circuit becomes overloaded. This is to prevent possible damage to electrical components, or the possible risk of fire.

- **Before renewing a fuse, always make sure that the ignition and the relevant circuit are switched off.**

The fuses are housed in a plastic box located under the bonnet at the rear right-hand side of the engine compartment (when viewed from the driver's seat).

The fuses are numbered to identify the circuit which they protect, and the circuits are represented by symbols on the plastic fusebox cover.

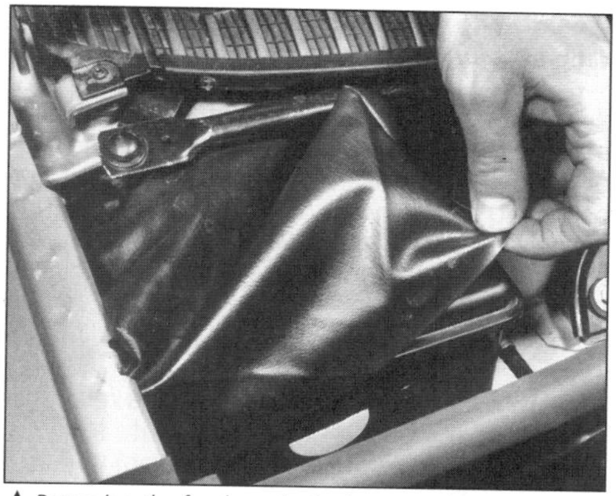

▲ *Removing the fusebox plastic sheet – early model*

▲ *Removing a fuse – early model*

▲ *Removing the fusebox lid securing clip – early model*

▲ *Fusebox cover removed to expose fuses and relays – early model*

To check a fuse, on early models, remove the plastic sheet from the fusebox lid, then remove the metal spring clip, and open the cover. On later models, release the tab at the front of the fusebox cover, and open the cover. All fuses are a push fit in their sockets, and a blown fuse can be recognised by a break in the silver 'inverted V' connection between the two terminals.

If a fuse has blown, renew it with one of identical rating. The fuses are colour-coded to show their rating as follows:

Colour	Rating [amps]
Red	10
Blue	15
Yellow	20
Clear	25
Green	30

Spare fuses are located in the fusebox cover.

Always use proper fuses of the correct type and rating, and **never** use wire or any other material to bridge the gap where a fuse should be.

Note that the same fuse should never be renewed more than once without investigating the source of the trouble. Seek advice from someone suitably qualified if necessary.

Various additional fuses may be used depending on the equipment fitted (eg radio, anti-lock braking system, engine management system), and these may be located in various positions depending on model. If a problem is suspected which may be due to a fuse not located in the fusebox, it's best to seek advice from a Ford dealer, who will know where on the car to find the relevant components.

RELAYS

Various types of relays are used in some circuits, and a faulty relay may prevent one or more components from working.

- **Before renewing a relay, always make sure that the ignition and the relevant circuit are switched off.**

Relays are of the plug-in type, and most of the major relays are located in the fusebox, with symbols on the fusebox cover representing the circuit covered by each relay.

Various additional relays may be used depending on the equipment fitted, and these may be located in various positions depending on model. If a problem is suspected which may be due to a relay not located in the fusebox, it's best to seek advice from a Ford dealer, who will know where on the car to find the relevant components.

To remove a relay, simply pull it from its socket.

It's difficult to tell visually whether a relay is faulty, and the best test is to substitute a known good relay to see if the relevant circuit then operates – if it does, then it's probably the relay which was causing the problem.

The following information is intended as a guide to enable you to spot some of the more obvious faults which may cause your car to fail the MOT test.

Obviously it isn't possible to check the car to the same standard as a professional MOT tester, who will be highly experienced and will have all the necessary tools and equipment.

Although we can't cover all the points of the test, the following should provide you with a good indication as to the general condition of the car, and will enable you to identify any obvious problem areas before submitting your car for the test.

Further explanation of most of the checks, and details of how to cure any problems discovered can be found in the relevant *Owners Workshop Manual* for your particular car (OWM 903 for Sierra 4-cylinder engine models, or OWM 904 for Sierra V6 engine models).

Lights

- All lights must work reliably, and the lenses and reflectors must not be damaged.
- Pairs of similar lights must be of the same brightness (eg both rear lights).
- Both headlights must show the same colour, and must be correctly aimed so that they light up the road adequately, but don't dazzle other drivers.
- The brake lights must work when the brake pedal is pressed.
- The direction indicator lights must flash between one and two times per second.
- All switches and driver's tell-tale lights must be fitted, secure and in good working order.
- No lamp should be adversely affected by the operation of any other lamp.

Steering and suspension

- All the components and their mounting points must be secure, and there must be no damage or excessive corrosion.
- There must be no excessive free play in the joints and bushes, and all rubber gaiters must be secure and undamaged.
- The steering mechanism must operate smoothly without excessive free play or roughness.
- There should be no excessive free play or roughness in the wheel bearings, but there should be sufficient free play to prevent tightness or binding.
- There should be no major fluid leaks from the shock absorbers and, when each corner of the car is depressed, it should rise and then settle in its normal position.
- The driveshafts and propeller shaft (on rear wheel drive cars) must not be damaged or distorted.

Brakes

- It should be possible to operate the handbrake without excessive force or excessive movement of the lever, and it must not be possible for the lever to release unintentionally. The handbrake must be able to lock the relevant wheels on which it operates.
- The brake pedal must not be damaged, and when the pedal is pressed, resistance should be felt near the top of its travel – hard resistance should be felt, and the pedal should not move down towards the floor. The resistance should be firm and not spongy.
- There should be no signs of any fluid leaks anywhere in the braking system.
- The wheels should turn freely when the brakes are not applied.
- When the brakes are applied, the brakes on all four wheels must work, and the car must stop evenly in a straight line without pulling to one side.

- All the braking system components must be secure, and there must be no signs of excessive wear or corrosion.

Tyres and wheels

- The tyres must be in good condition, and there must be no signs of excessive wear or damage (refer to *'Regular checks'* on page 76).
- Tyres at the same end of the car must be of the same size and type. Always fit radial tyres; crossply tyres are not suitable for modern cars.

Seatbelts

- The mountings must be secure, and must not be loose or excessively corroded.
- The seatbelt webbing should not be frayed or damaged (this must be checked along the full length of the belts).
- When a seatbelt is fastened, the locking mechanism should hold securely and should release when intended.
- In the case of inertia reel seatbelts, the retractors should work properly when the belts are released.

General

- The windscreen wipers and washers must work properly, and the wiper blades must clear the windscreen without smearing.
- The swept area of the windscreen must have a zone 290mm wide, centred on the steering wheel, free of impact damage greater than 10mm across; this zone must be not encroached by any stickers by more than 10mm. Similar restrictions apply to the remainder of the swept area. Here, the damage or obstruction must not exceed 40mm across.
- The horn must work properly, be loud enough to be heard by other road users and emit a constant tone. Two-tone horns, which alternate between the two frequencies (i.e. they warble), are illegal.
- The exhaust mountings must be secure, and the system itself must be free from leaks and serious corrosion.
- There must be no serious corrosion or damage to the vehicle structure. Corrosion of body panels will not necessarily cause a car to fail the test, but there should be no serious corrosion of any of the 'load bearing' structural components.
- Door handles and locks should work properly from both inside and outside; boot lids and tailgates must close securely; seats and spare tyres must be securely mounted; petrol caps must seal and fasten properly.
- Registration plates must be securely fitted, front and rear, clearly legible, not obscured by items such as a tow bar, with the letters and figures correctly formed and spaced. (Letters or figures formed or positioned so that they are likely to be mis-read, i.e. to personalise a number plate, are now illegal – the vehicle will be failed.)
- Vehicles must display a legible Vehicle Identification Number (VIN). This is either a plate secured to the body or chassis in the engine compartment or a number stamped or etched on the vehicle.
- The exhaust gas emissions level must be within certain limits depending on the age of the car (special equipment is required to measure exhaust gas emissions), and the MOT test also includes a visual check for excessive exhaust smoke when the engine is running. Generally, the car should meet the emission regulations if regular servicing has been carried out.

The following table isn't intended as an exhaustive guide to fault finding, but it summarises some of the more common faults which may crop up during a car's life.

Hopefully, the table should help you to find the cause of a problem, even if you can't cure it yourself.

When confronted with a fault, try to think calmly and logically about the symptom(s), and you should be able to work out what the fault can't be! Check one item at a time, otherwise if you do clear the fault, you may not know what was causing it.

The commonest cause of difficulty is starting, especially in the winter. Make sure that your battery is kept fully charged, and that the ignition components (ignition coil, distributor cap, spark plugs, HT leads, etc) are in good condition, clean and dry (refer to *'Servicing'* on page 80).

Further details of fault diagnosis, along with comprehensive renewal procedures for most components can be found in our *Owners Workshop Manual* for your particular model (OWM 903 for Sierra 4-cylinder engine models, or OWM 904 for Sierra V6 engine models).

SYMPTOM	POSSIBLE CAUSES
ENGINE	
Starter motor doesn't turn, and headlights don't come on	● Flat battery ● Loose, dirty or corroded battery connections ● Other electrical or wiring fault
Starter motor doesn't turn, and headlights dim	● Battery charge low ● Loose, dirty or corroded battery connections ● Faulty starter motor ● Other electrical or wiring fault ● Seized engine
Starter motor doesn't turn, and headlights are bright	● Loose or dirty starter motor connections ● Faulty starter motor or ignition switch ● Gear selector lever not in position **N** or **P** (automatic transmission only) ● Other electrical or wiring fault
Starter motor spins but doesn't turn engine	● Faulty starter motor ● Engine fault
Engine turns slowly but won't start	● Battery charge low ● Electrical or wiring fault ● Wrong grade of engine oil (refer to *'Service specifications'* on page 68)

SYMPTOM	POSSIBLE CAUSES
ENGINE (continued)	
Engine turns but won't fire, or engine starts but won't keep running	● Fuel tank empty ● Damp or dirty ignition system components ● Dirty or loose ignition system connections ● Incorrectly adjusted or faulty spark plugs (refer to *'Servicing'* on page 100) ● Fuel system fault ● Other electrical or wiring fault
Engine idles, but stalls when accelerator pedal is depressed	● Blocked or dirty air cleaner filter element (refer to *'Servicing'* on page 105) ● Fuel system fault ● Ignition system fault ● Electrical fault
Poor acceleration, misfiring or lack of power	● Incorrectly adjusted or faulty spark plugs (refer to *'Servicing'* on page 100) ● Incorrect engine valve clearances (OHC and V6 engines only) ● Fuel system fault ● Ignition system fault ● Electrical fault
Engine continues to run when ignition is switched off	● Engine overheating (possibly due to low coolant level – refer to *'Regular checks'* on page 75) ● Incorrectly adjusted spark plugs, or wrong type of spark plugs fitted (refer to *'Servicing'* on page 100) ● Wrong grade of petrol ● Fuel system fault ● Excessive build-up of carbon inside engine ('decoke' required) ● Faulty ignition switch ● Other electrical or wiring fault
Engine doesn't reach normal operating temperature	● Faulty cooling system thermostat ● Faulty temperature gauge or sensor

SYMPTOM	POSSIBLE CAUSES
ENGINE (continued)	
Engine overheats, or temperature gauge reads too high	● Airflow to radiator obstructed ● Blocked radiator, hose or engine coolant passage ● Coolant or engine oil level low (refer to *'Regular checks'* on page 74) ● Coolant hose(s) leaking, worn or damaged ● Faulty cooling system thermostat ● Faulty water pump ● Faulty water pump drive ● Faulty temperature gauge or sensor ● Faulty electric cooling fan or switch (where applicable)
Ignition warning light comes on when engine is running	● Alternator drivebelt loose or broken (refer to *'Servicing'* on page 102) ● Faulty alternator ● Other electrical or wiring fault
Oil pressure warning light comes on when engine is running	● Oil level below 'MIN' mark on dipstick – refer to *'Regular checks'* on page 74 (if oil level is correct, and light is still on, seek advice, but **don't** start the engine) ● Oil leak ● Faulty oil pressure switch ● Badly worn engine components
GEARBOX, TRANSMISSION AND CLUTCH	
Difficulty in engaging gear (manual gearbox)	● Worn or faulty gearbox components ● Worn or faulty clutch mechanism ● Engine idle speed too high
Clutch slips (car does not accelerate when engine revs increase – manual gearbox)	● Faulty clutch mechanism ● Contaminated or worn clutch
Car gearchange is harsh (automatic transmission)	● Fault in transmission
Car pull away/gear change sluggish (automatic transmission)	● Low transmission fluid level (refer to *'Servicing'* on page 101) ● Fault in transmission

SYMPTOM	POSSIBLE CAUSES
BRAKES	
Brakes feel 'spongy'	● Air in brake hydraulic system ● Fluid leak in brake system
Excessive brake pedal travel	● Faulty rear brake mechanism ● Air in brake hydraulic system ● Fluid leak in brake system
Brakes require excessive pedal pressure	● Damp, dirty or contaminated brake components ● Fluid leak in brake system ● Faulty or seized brake components ● Faulty brake servo
Car pulls to one side	● Incorrect tyre pressure(s) or uneven tyre wear (refer to *'Regular checks'* on page 76) ● Faulty or seized brake components ● Worn or contaminated brake friction material ● Incorrect wheel alignment ● Worn or faulty steering components ● Worn or faulty suspension components
Brakes squeal and/or judder	● Badly worn or corroded brake components ● Incorrectly assembled brake components ● Contaminated brake friction material ● Worn or faulty suspension components ● Worn or faulty steering components
Brake fluid level warning light comes on	● Low brake fluid level (refer to *'Regular checks'* on page 76) ● Faulty fluid level sensor or wiring
SUSPENSION AND STEERING	
Car becomes heavy to steer	● Low tyre pressure(s) (refer to *'Regular checks'* on page 76) ● Incorrect wheel alignment ● Worn or faulty steering components ● Worn or faulty suspension components
Car pulls to one side	● Refer to *'Brakes'* section of table

SYMPTOM	POSSIBLE CAUSES
SUSPENSION AND STEERING (continued)	
Car wanders	● Incorrect tyre pressure(s) or uneven tyre wear (refer to *'Regular checks'* on page 76) ● Car unevenly loaded ● Incorrect wheel alignment ● Worn or faulty suspension components ● Worn or faulty steering components
Car vibrates when driving	● Loose wheel bolt(s) ● Wheel(s) out of balance ● Worn or damaged driveshaft ● Worn or faulty suspension components ● Worn or faulty steering components ● Worn or faulty brake components
Hard or choppy ride	● Incorrect tyre pressure(s) (refer to *'Regular checks'* on page 76) ● Worn or faulty suspension components
Car leans excessively when cornering	● Roof rack overloaded ● Car unevenly loaded ● Worn or faulty suspension components
Uneven tyre wear	● Incorrect tyre pressure(s) (refer to *'Regular checks'* on page 76) ● Incorrect wheel alignment ● Wheel(s) out of balance ● Worn or faulty suspension components ● Worn or faulty steering components ● Brakes 'grabbing'
ELECTRICS	
Electrical systems don't work	● Loose, dirty or corroded battery connections ● Faulty earth connection ● Battery charge low ● Blown fuse (refer to *'Bulb, fuse and relay renewal'* on page 126) ● Faulty relay ● Fuse link, connecting main wiring loom to battery blown (seek advice)

SYMPTOM	POSSIBLE CAUSES
ELECTRICS (continued)	
Direction indicators don't work	● Blown fuse (refer to *'Bulb, fuse and relay renewal'* on page 126) ● Faulty earth connection ● Faulty direction indicator relay ● Faulty direction indicator switch
Bulbs burn out repeatedly	● Faulty connections at light socket ● Faulty alternator regulator
All lights dim when engine speed drops to idle	● Loose alternator drivebelt (refer to *'Servicing'* on page 102) ● Battery charge low ● Faulty alternator

This Section should help you to understand some of the odd words of phrases used at garages and by 'car enthusiasts' which you may not be familiar with.

A

ABS – Abbreviation for Anti-lock Braking System. Uses sensors at each wheel to sense when the wheels are about to lock, and releases the brakes to prevent locking. This process occurs many times per second, and allows the driver to maintain steering control when braking hard.
Accelerator pump – A device attached to many *carburettors* which provides a spurt of extra fuel to the carburettor fuel/air mixture when the accelerator pedal is suddenly pressed down.
Additives – Compounds which are added to petrol and oil to improve their quality and performance.
Advance and retard – A system for altering the *ignition timing*.
AF – An abbreviation of 'Across Flats', the way in which many nuts, bolts and spanners are identified. AF is usually preceded by an Imperial unit of measurement – eg ½in AF. Unless otherwise stated, all metric measurements are assumed to be AF, so the abbreviation is not normally used for metric nuts, bolts and spanners.
Air cooling – Alternative method of engine cooling in which no water is used. An engine-driven fan forces air at high speed over the surfaces of the engine.
ALB – Abbreviation for Anti-Lock Braking System. See *'ABS'*.
Alternator – A device for converting rotating mechanical energy into electrical energy. In modern cars, it has superseded the dynamo for charging the battery because of its much greater efficiency.
Ammeter – A device for measuring electrical current – the current supplied to the battery by the alternator, or drawn from the battery by the car's electrical systems.
Antifreeze – A chemical mixed with the water in the cooling system to lower the temperature at which the coolant freezes, and in modern cars to prevent corrosion of the metal in the cooling system.
Anti-roll bar – A metal bar mounted transversely across the car, connecting the two sides of the suspension, which counteracts the natural tendency for the car to lean when cornering.
Aquaplaning – A word used to describe the action of a tyre skating across the surface of water.
Automatic transmission – A type of *gearbox* which selects the correct gear ratio automatically according to engine speed and load.
Axle – Spindle on which a wheel revolves.

B

Balljoint – A ball-and-socket type joint used in steering and *suspension* systems, which allows relative movement in more than one plane.
Battery condition indicator – A device for measuring electrical voltage (a voltmeter) connected via the ignition switch to the car battery. Unlike an *ammeter*, it gives an indication that a battery is close to failing. **Also**, most 'maintenance-free' batteries have a battery condition indicator fitted to their casing, which consists of a small disc which changes its colour when the battery is close to failure and requires renewal.
Bearing – Metal or other hard wearing surface against which another part moves, and which is designed to reduce friction and wear (bearings are usually lubricated).

Bendix drive – A device on some types of starter motor which allows the motor to drive the engine for starting, then disengages when the engine starts to run.
BHP – see *Horsepower*.
Big end – The end of a *connecting rod* which is attached to the *crankshaft*. It incorporates a *bearing* and transmits the linear movement of the connecting rod to the crankshaft.
Bleed nipple (or valve) – A hollow screw which allows air or fluid to be bled out of a system when it is loosened.
Brake caliper – The part of a *disc brake* system which houses the *brake pads* and the hydraulically-operated pistons.
Brake disc – A rotating disc, coupled to a roadwheel, which is clamped between hydraulically operated friction pads in a *disc brake* system.
Brake fade – A temporary loss of braking efficiency due to overheating of the brake friction material.
Brake pad – The part of a *disc brake* system which consists of the friction material and a metal backing plate.
Brake shoe – The part of a *drum brake* system which consists of the friction material and a curved metal former.
Breather – A device which allows fresh air into a system or allows contaminated air out.
Bucket tappet – A bucket shaped component used in some engines to transfer the rotary movement of the *camshaft* to the up-and-down movement required for *valve* operation.
Bump stop – A hard piece of rubber used in many *suspension* systems to prevent the moving parts from contacting the body during violent suspension movements.

C

Camber angle – The angle at which the front wheels are set from the vertical, when viewed from the front of the car. Positive camber is the amount in degrees which the wheels are tilted out at the top.
Cam follower – A piece of metal used to transfer the rotary movement of the *camshaft* to the up-and-down movement required for *valve* operation.
Camshaft – A rotating shaft driven from the *crankshaft* with lobes or cams used to operate the engine *valves* via the *valve gear*.
Carbon leads – Ignition HT leads incorporating carbon (black fibres) which eliminates the need for separate radio and TV *suppressors*.
Carburettor – A device which is used to mix air and fuel in the proportions required for burning by the engine under all conditions of engine running.
Castor angle – The angle between the front wheels pivot points and a vertical line when viewed from the side of the car. Positive castor is when the axis is inclined rearwards.
Catalytic converter – A device incorporated in the exhaust system which speeds up the natural decomposition of the exhaust gases, and reduces the amount of harmful gases released into the atmosphere. Cars fitted with catalytic converters must be operated on *unleaded petrol*, as *leaded petrol* will destroy the catalyst.
Centrifugal advance – System of ignition *advance and retard* incorporated in many *distributors* in which weights rotating on a shaft alter the ignition timing according to engine speed.
Choke – This has two common meanings. It is used to describe the device which shuts off some of the air in a *carburettor* during cold starting (in order to provide extra fuel), and it may be manually or automatically operated. It's also used as a general term to describe a carburettor throttle bore.
Clutch – A friction device which allows two rotating components to be coupled together smoothly, without the need for either rotating component to stop moving.
Coil spring – A spiral coil of spring steel used in many *suspension* systems.
Combustion chamber – Shaped area in the *cylinder head* into which the fuel/air mixture is compressed by the *piston* and where the spark from the *spark plug* ignites the mixture.
Compression ratio (CR) – A term used to describe the amount by which the fuel/air mixture is compressed as a *piston* moves from the bottom to the top of its travel, and expressed as a number. For example an 8.5:1 compression ratio means that the volume of fuel/air mixture above the piston when the piston is at the bottom of its stroke is 8.5 times

that when the piston is at the top of its stroke.
Compression tester – A special type of pressure gauge which can be screwed into a *spark plug* hole, which measures the pressure in the cylinder when the engine is turning but not firing. This gives an indication of engine wear or possible leaks.
Condenser (capacitor) – A device in a *contact breaker point distributor* which stores electrical energy and prevents excessive sparking at the contact breaker points.
Connecting rod ('con-rod') – Metal rod in the engine connecting a *piston* to the *crankshaft*.
Constant velocity (CV) joint – A joint used in *driveshafts*, where the instantaneous speed of the input shaft is exactly the same as the instantaneous speed of the output shaft at any angle of rotation. This does not occur in ordinary *universal joints*.
Contact breaker points – A device in the *distributor* which consists of two electrical points (or contacts) and a cam which opens and closes them to operate the *HT* electrical circuit which provides the spark at the *spark plugs*.
Crankcase – The area of the *cylinder block* below the *pistons* which houses the *crankshaft*.
Crankshaft – A cranked shaft which is driven by the *pistons* and provides the engine output to the *transmission*.
Crossflow cylinder head – A *cylinder head* in which the inlet and exhaust *valves* and *manifolds* are on opposite sides.
Crossply tyre – A tyre whose construction is such that the weave of the fabric material layers is running diagonally in alternately opposite directions to a line around the circumference of the tyre.
Cubic capacity – The total volume within the *cylinders* of an engine which is swept by the movement of the *pistons*.
CVH – A term applied by the Ford Motor Company to their overhead camshaft engines which incorporate a hemispherical *combustion chamber*. CVH means Compound Valve angle, Hemispherical combustion chamber.
Cylinder – Close fitting metal tube in which a *piston* slides. In the case of an engine, the cylinders may be bored directly into the *cylinder block*, or on some engines, cylinder liners are used which rest in the cylinder block and can be replaced when worn with matching pistons to avoid the requirement for *reboring* the cylinder block.
Cylinder block – The main engine casting which contains the *cylinders*, *crankshaft* and *pistons*.
Cylinder head – The casting at the top of the engine which contains the *valves* and associated operating components.

D

Damper – See *shock absorber*.
Dashpot – An oil-filled *cylinder* and *piston* used as a damping device in SU and Zenith/Stromberg CD type carburettors.
Dead axle (beam axle) – The simplest form of axle, consisting of a horizontal member attached to the car underbody by springs. This arrangement is used for the rear axle on some front-wheel-drive cars.
Decarbonising ('decoking') – Removal of all carbon deposits from the *combustion chambers* and the tops of the *pistons* and *cylinders* in an engine.
De Dion axle – A rear axle consisting of a cranked tube attached to the wheel hubs, with a separately mounted *differential* gear and *driveshafts*. *Suspension* is normally through *coil springs* between the wheel hubs and car underbody.
Derv – Abbreviation for Diesel-Engined Road Vehicle. A term often used to refer to Diesel fuel.
Diaphragm – A flexible membrane used in some components such as fuel pumps. The diaphragm spring used on *clutches* is similar, but is made from spring steel.
Diesel engine – An engine which relies on the heat generated when compressing air to ignite the fuel, and which therefore doesn't need an *ignition system*. Diesel engines have much higher *compression ratios* than petrol engines, normally around 20:1.
Differential – A system of gears (generally known as a crownwheel and pinion) which allows the *torque* provided by the engine to be applied to both driving wheels. The differential divides the torque proportionally between the driving wheels to allow one wheel to turn faster than the other, for example during cornering.

DIN – This stands for Deutsche Industrie Norm (German Industry Standard), which provides international standards for measuring engine power, torque, etc.
Disc brake – A brake assembly where a rotating disc is clamped between hydraulically operated friction pads.
Distributor – A device used to distribute the *HT* current to the individual *spark plugs*. The distributor may also contain the *advance and retard* mechanism. On some older cars, the distributor also contains the *contact breaker points* assembly.
Distributor cap – Plastic cap which fits on top of the *distributor* and contains electrodes, in which the *rotor arm* rotates to distribute the *HT* spark voltage to the correct *spark plug*.
DOHC – Abbreviation for Double Overhead Camshaft (see *'Twin-cam'*).
Doughnut – A term used to describe the flexible rubber coupling used in some *driveshafts*.
Driveshaft – Term usually used to describe the shaft (normally incorporating *universal* or *constant velocity joints*), which transmits drive from a *differential* to one wheel. More commonly found in front-wheel-drive cars.
Drive train – A collective term used to describe the *clutch/gearbox/transmission* and the other components used to transmit drive to the wheels.
Drum brake – A brake assembly with friction linings on 'shoes' running inside a cylindrical drum attached to the wheel.
Dual circuit brakes – A *hydraulic* braking system consisting of two separate fluid circuits, so that if one circuit becomes inoperative, braking power is still available from the other circuit.
Dwell angle – A measurement which corresponds to the number of degrees of *distributor* shaft rotation during which the *contact breaker points* are closed during the ignition cycle of one *cylinder*. The angle is altered by adjusting the contact breaker points gap.

E

Earth strap – A flexible electrical connection between the battery and a car earth point, or between the engine/*gearbox* and the car body to provide a return current path flow to the battery.
EFI – Abbreviation for Electronic *Fuel Injection*.
Electrode – An electrical terminal, eg in a *spark plug* or *distributor cap*.
Electrolyte – A current-conducting solution inside the battery (consisting of water and sulphuric acid in the case of a car battery).
Electronic ignition – An *ignition system* incorporating electronic components in place of *contact breaker points*, which can produce a much higher spark voltage than a contact breaker system, and is less affected by worn components.
Emission control – The reduction or prevention of the release into the atmosphere of poisonous fumes and gases from the engine and fuel system of a car. Required to different degrees by the laws of some countries, and achieved by engine design and the use of special devices and systems.
Epicyclic gears (planetary gears) – A gear system used in many *automatic transmissions* where there is a central 'sun' gear around which smaller 'planet' gears rotate.
Exhaust gas analyser – An instrument used to measure the amount of pollutants (mainly carbon monoxide) in a car's exhaust gases.
Expansion tank – A container used in many cooling systems to collect the overflow from the car's cooling system as the coolant heats up and expands.

F

Filter – A device for removing foreign particles from air, fuel or oil.
Final drive – A collective term (often expressed as a gear ratio) for the crownwheel and pinion (see *Differential*).
Flat engine – Form of engine design in which the *cylinders* are opposed horizontally, usually with an equal number on each side of the central *crankshaft*.
Float chamber – The part of a *carburettor* which contains a float and *needle valve* for

controlling the fuel level in the reservoir.
Flywheel – A heavy rotating metal disc attached to the *crankshaft* and used to smooth out the pulsing from the *pistons*.
Four stroke (cycle) – A term used to describe the four operating strokes of a *piston* in a conventional car engine. These are (1) Induction – drawing the air/fuel mixture into the engine as the piston moves down; (2) Compression – of the fuel/air mixture as the piston rises; (3) Power stroke – where the piston is forced down after the fuel/air mixture has been ignited by the *spark plug*, and (4) Exhaust stroke – where the piston rises and pushes the burnt gases out of the *cylinder*. During these operations, the inlet and exhaust *valves* are opened and closed at the correct moment to allow the fuel/air mixture in, the exhaust gases out, or to provide a gas-tight compression chamber.
Fuel injection – A method of injecting fuel into an engine. Used in *Diesel engines* and also on some petrol engines in place of a *carburettor*.
Fuel injector – Device used on *fuel injection* engines to inject fuel directly or indirectly into the *combustion chamber*. Some fuel injection systems use a single fuel injector, while some systems use one fuel injector for each *cylinder* of the engine.

G

Gasket – Compressible material used between two surfaces to provide a leakproof joint.
Gearbox – A group of gears and shafts installed in a housing, positioned between the *clutch* and the *differential*, and used to keep the engine within its safe operating speed range as the speed of the car changes.

H

Half-shaft – A *driveshaft* used to transmit the drive from the *differential* to one of the rear wheels.
Hardy-Spicer joint (Hooke's or Cardan joint) – See *Universal joint*.
Helical gears – Gears in which the teeth are cut at an angle across the circumference of the gear to give a smoother mesh between gears and quieter running.
Horsepower – A measurement of power. Brake Horsepower (BHP) is a measure of the power required to stop a moving body.
HT – Abbreviation of High Tension (meaning high voltage) used to describe the *spark plug* voltage in an *ignition system*.
Hub carrier – A component usually found at each front corner of a car which carries the wheel and brake assembly, and to which the *suspension* and steering components are attached.
Hydraulic – A term used to describe the operation of a system by means of fluid pressure.
Hypoid gear – A gear with curved teeth which transmits drive through a right-angle, where the centreline of the drive gear is offset from the centreline of the driven gear. The meshing action of hypoid gears allows a larger and therefore stronger drive gear, and the meshing noise is reduced in comparison with conventional gears.

I

Independent suspension – A *suspension* system where movement of one wheel has no effect on the movement of the other, eg independent front suspension.
Ignition coil – An electrical coil which forms part of the *ignition system* and which generates the *HT* voltage.
Ignition system – The electrical system which provides the spark to ignite the air/fuel mixture in the engine. Normally the system consists of the battery, *ignition coil*, *distributor*, ignition switch, *spark plugs* and wiring.
Ignition timing – The time in the *cylinder* firing cycle at which the ignition spark (provided by the *spark plug*) occurs. The spark timing is normally a few degrees of *crankshaft* rotation before the *piston* reaches the top of its stroke, and is expressed as a number of degrees before top-dead-centre (BTDC).
Inertia reel – Automatic type of seat belt mechanism which allows the wearer to move freely in normal use, but which locks on sensing either sudden deceleration or a sudden movement of the wearer.

In-line engine – An engine in which the *cylinders* are positioned in one row as opposed to being in a *flat* or *vee* configuration.

J

Jet – A calibrated nozzle or orifice in a *carburettor* through which fuel is drawn for mixing with air.
Jump leads – Heavy electric cables fitted with clips to enable a car's battery to be connected to another battery for emergency starting.

K

Kerb weight – The weight of a car, unladen but ready to be driven, ie with enough fuel, oil, etc, to travel an arbitrary distance.
Kickdown – A device used on *automatic transmissions* which allows a lower gear to be selected for improved acceleration by fully depressing the accelerator.
Kingpin – A device which allows the front wheel of a car to swivel about a near vertical axis.
Knocking – See *'Pinking'*.

L

Laminated windscreen – A windscreen which has a thin plastic layer sandwiched between two layers of toughened glass. It will not shatter or craze when hit.
Lead-free petrol – Contains no lead. It has no lead added during manufacture, and the natural lead content is refined out. This type of petrol is not currently available for general use in the UK, and should not be confused with *unleaded* petrol.
Leaded petrol – Normal 4-star petrol. Has a low amount of lead added during manufacture, in addition to the natural lead found in crude oil.
Leading shoe – A *drum brake* shoe of which the leading end (the one moved by the operating *pistons*) is reached first by a given point on the drum during normal forward rotation. A simple drum brake will have one leading and one trailing (the opposite) shoe.
Leaf spring – A spring commonly used on cars with a *live axle*, consisting of several long curved steel plates clamped together.
Limited slip differential – A type of *differential* which prevents one wheel from standing still while the other wheel spins excessively. Often used on high-performance cars.
Live axle – An axle through which power is transmitted to the rear wheels.
Loom – A complete car wiring system or section of a wiring system consisting of all the wires of correct length, etc, to wire up the various circuits.
LT – Abbreviation of Low Tension (meaning low voltage), used to describe battery voltage in the *ignition system*.

M

MacPherson strut – An independent front *suspension* system where the swivelling, springing and shock absorbing action of the wheels is dealt with by a single assembly.
Manifold – A device used for ducting the air/fuel mixture to the engine (inlet manifold), or the exhaust gases from the engine (exhaust manifold).
Master cylinder – A *cylinder* containing a *piston* and *hydraulic* fluid, directly coupled to a foot pedal (eg brake or *clutch* master cylinder). Used for transmitting pressure to the brake or clutch operating mechanism.
Metallic paint – Paint finish incorporating minute particles of metal to give added lustre to the colour.
Multigrade – Lubricating oil whose *viscosity* covers that of several single grade oils, making it suitable for use over a wider range of operating conditions.

N

Needle bearing – Type of *bearing* in which needle or cone-shaped rollers are used around the circumference to reduce friction.
Needle valve – A component of the *carburettor* which restricts the flow of fuel or fuel/air mixture according to the position of the valve in an orifice or *jet*.

Negative earth – Electrical system (almost universally adopted) in which the negative terminal of the car battery is connected to the car body. The polarity of all the electrical equipment is determined by this.

O

Octane rating – A scale rating for grading petrol.
ohc (overhead cam) – Describes an engine in which the *camshaft* is situated above the *cylinder head*, and operates the *valve gear* directly.
ohv (overhead valve) – Describes an engine which has its *valves* in the *cylinder head*, but with the *valve gear* operated by *pushrods* from a *camshaft* situated lower in the engine.
Oil cooler – Small *radiator* fitted in the oil circuit and positioned in a cooling airflow to cool the oil. Used mainly on high-performance engines.
Overdrive – A device coupled to a car's *gearbox* which raises the output gear ratio above the normal 1:1 of top gear.
Oversteer – A tendency for a car to turn more tightly into a corner than intended.

P

PCV (Positive Crankcase Ventilation) – A system which allows fumes and vapours which build up in the *crankcase* to be drawn into the engine for burning.
Pinion – A gear with a small number of teeth which meshes with one having a larger number of teeth.
Pinking – A metallic noise from the engine often caused by the *ignition timing* being too far *advanced*. The noise is the result of pressure waves which cause the *cylinder* walls to vibrate when the ignited fuel/air mixture is compressed.
Piston – Cylindrical component which slides in a closely-fitting metal tube or *cylinder* and transmits pressure. The pistons in an engine, for example, compress the fuel/air mixture, transmit the power to the *crankshaft*, and push the burnt gases out through the exhaust *valves*.
Piston ring – Hardened metal ring which is a spring fit in a groove running round the *piston* to ensure a gas-tight seal between the piston and *cylinder* wall.
Positive earth – The opposite of *negative earth*.
Power steering – A steering system which uses *hydraulic* fluid pressure (provided by an engine-driven pump) to reduce the effort required to steer the car.
Pre-ignition – See *'Pinking'*.
Propeller shaft – The shaft which transmits the drive from the *gearbox* to the rear axle in a front-engined rear-wheel-drive car.
Pushrod – A rod which is moved up and down by the rotary motion of the *camshaft* and operates the *rocker arms* in an *ohv* engine.

Q

Quarter light – A triangular window mounted in front or behind the main front or rear windows, usually in the front door, or behind the rear door.
Quartz-halogen bulb – A bulb with a quartz envelope (instead of glass), filled with a halogen gas. Gives a brighter, more even spread of light than an ordinary bulb.

R

Rack and pinion – Simplest form of steering mechanism which uses a *pinion* gear to move a toothed rack.
Radial ply tyre – A tyre in which the fabric material plies are arranged laterally, at right angles to the circumference.
Radiator – Cooling device through which the engine coolant is passed, situated in an air flow and consisting of a system of fine tubes and fins for rapid heat dissipation.
Radius arms (rods) – Locating arms sometimes used with a *live axle* to positively locate it in the fore-and-aft direction.
Rebore – The process of enlarging the *cylinder* bores to a very accurately specified measurement in order to fit new *pistons* to overcome wear in the engine. Not normally necessary unless the engine has covered a very high mileage.

Recirculating ball steering – A derivative of *worm and nut* steering, where the steering shaft motion is transmitted to the steering linkage by balls running in the groove of a worm gear.
Rev counter – See *Tachometer*.
Rocker arm – A lever which rocks on a central pivot, with one end moved up and down by the *camshaft*, and the other end operating an inlet or exhaust *valve*.
Rotary engine – See '*Wankel engine*'.
Rotor arm – A rotating arm in the *distributor* which distributes the *HT* spark voltage to the correct *spark plug*.
Running on – A tendency for an engine to keep on running after the ignition has been switched off. Often caused by a badly maintained engine or the use of an incorrect grade of fuel.

S

SAE – Society of Automotive Engineers (of America). Lays down international standards for the classification of engine performance and many other specifications, but is most commonly used to classify oils.
Safety rim – A special wheel rim shape which prevents a deflated tyre from rolling off the wheel.
Sealed beam – A sealed headlamp unit where the filament is an integral part and cannot be renewed separately.
Semi-trailing arm – A common form of independent rear *suspension*.
Servo – A device for increasing the normal effort applied to a control.
Shock absorber – A device for damping out the up-and-down movement of the *suspension* when the car hits a bump in the road.
Spark plug – A device with two *electrodes* insulated from each other by a ceramic material, which screws into an engine *combustion chamber*. When the *HT* voltage is applied to the plug terminal, a spark jumps across the electrodes and ignites the fuel/air mixture.
Squab – Another name for a seat cushion.
Steel-braced tyre – Tyre in which extra plies containing steel cords are incorporated with the fabric plies to give added strength.
Steering arm (knuckle) – A short arm on the front *hub carrier* to which the steering linkage connects.
Steering gear – A general term used to describe the steering components, usually refers to a steering rack-and-pinion assembly.
Steering rack – See *Rack and pinion*.
Stroboscopic light – A light switched on and off by the engine *ignition system* which is used for checking the *ignition timing* when the engine is running.
Stroke – The total distance travelled by a single *piston* in its *cylinder*.
Stub axle – A short axle which carries one wheel.
Subframe – A small frame which is mounted on the car's body, and carries the *suspension* and/or the *drivetrain* assemblies.
Sump – The main reservoir for the engine oil.
Supercharger – A device which uses an engine-driven turbine (usually driven by a belt or gears from the *crankshaft*) to drive a compressor which forces air into the engine, providing increased fuel/air mixture flow, and therefore increased engine efficiency. Sometimes used on high-performance engines.
Suppressor – A device which is used to reduce or eliminate electrical interference caused by the *ignition system* or other electrical components.
Suspension – A general term used to describe the components which suspend the car body on its wheels.
Swing axle – A *suspension* arm which is pivoted near the front-to-rear centreline of the car, and which allows the wheel to swing vertically about that pivot point.
Synchromesh – A device in a *gearbox* which synchronises the speed of one gear shaft with another to produce smooth, noiseless engagement of the gears.

T

Tachometer – Also known as a rev counter, indicates engine speed in revolutions per minute (rpm).
Tappet – A term often used to refer to the component which transmits the rotary *camshaft* movement to the up-and-down movement required for *valve* operation.

Thermostat – A device which is sensitive to changes in engine coolant temperature, and opens up an additional path for coolant to flow through the *radiator* (to increase the cooling) when the engine has warmed up.
Tie-rod – A rod which connects the *steering arms* to the *steering gear*.
Timing belt – Fabric or rubber belt engaging on sprocket wheels and driving the *camshaft* from the *crankshaft*.
Timing chain – Metal flexible link chain engaging on sprocket wheels and driving the *camshaft* from the *crankshaft*.
Timing marks – Marks normally found on the *crankshaft* pulley or the *flywheel* and used for setting the ignition firing point with respect to a particular *piston*.
Toe-in/toe-out – The amount by which the front wheels point inwards or outwards from the straight-ahead position when steering straight ahead.
Top Dead Centre (TDC) – The point at which a *piston* is at the top of its *stroke*.
Torque – The turning force generated by a rotating component.
Torque converter – A coupling where the driving *torque* is transmitted through oil. At low speeds there is very little transfer of torque from the input to the output. As the speed of the input shaft increases, the direction of fluid flow within a system of vanes changes, and torque from the input impeller is transferred to the output turbine. The higher the input speed, the closer the output speed approaches it, until they are virtually the same.
Torsion bar – A metal bar which twists about its own axis, and is used in some *suspension* systems.
Toughened windscreen – A windscreen which when hit, will shatter in a particular way to produce blunt-edged fragments or will craze over but remain intact. A zone toughened windscreen has a zone in front of the driver which crazes into larger parts to reduce the loss of visibility which occurs when toughened windscreens break, but is otherwise similar.
Track rod – See *Tie-rod*.
Trailing arm – A form of independent *suspension* where the wheel is attached to a swinging arm, and is mounted to the rear of the arm pivot.
Transaxle – A combined *gearbox*/axle assembly from which two *driveshafts* transmit the drive to the wheels.
Transmission – A general term used to describe some or all of the *drivetrain* components excluding the engine, most commonly used to describe automatic gearboxes.
Turbocharger – A device which uses a turbine driven by the engine exhaust gases to drive a compressor which forces air into the engine, providing increased fuel/air mixture flow, and therefore increased engine efficiency. Commonly used on high-performance engines.
Twin-cam – Abbreviation for twin overhead *camshafts* (see *'ohc'*). Used on engines with a *crossflow cylinder head*, usually with one camshaft operating the inlet *valves* and the other operating the exhaust valves. Gives improved engine efficiency due to improved fuel/air mixture and exhaust gas flow in the *combustion chambers*.
Two stroke (cycle) – A common term used to describe the operation of an engine where each downward *piston* stroke is a power stroke. The fuel/air mixture is directed to the crankcase where it's compressed by the descending piston and pumped into the *combustion chamber*. As the piston rises, the mixture is compressed and ignited, which forces the piston down. The burnt gases flow from the exhaust port, but the piston is now compressing another fuel/air mixture charge in the crankcase which repeats the cycle.

U

Understeer – A tendency for a car to go straight on when turned into a corner.
Universal joint – A joint that can swivel in any direction whilst at the same time transmitting *torque*. This type of joint is commonly used in *propeller shafts* and some *driveshafts*, but is not suitable for some applications because the input and output shaft speeds are not the same at all positions of angular rotation. The type in common use is known as a Hardy-Spicer, Hooke's or Cardan joint.
Unleaded petrol – Has no lead added during manufacture, but still has the natural lead content of crude oil. Generally available in the

UK, most modern cars can use this type of petrol, but seek advice first, as engine adjustments may be required. Engine damage can occur if unleaded petrol is used incorrectly. Not to be confused with *lead-free* petrol which is not currently available in the UK.
Unsprung weight – The part of the car which is not supported by the springs.

V

Vacuum advance – System of ignition *advance and retard* used in some *distributors* where the vacuum in the engine inlet *manifold* is used to act on a *diaphragm* which alters the *ignition timing* as the vacuum changes due to the throttle position.
Valve – A device which opens or closes to allow or stop gas or fluid flow.
Valve gear – A general term used for the components which are acted on by the *camshaft* in order to operate the *valves*.
16-valve – Term used to describe a four *cylinder* engine with four *valves* per cylinder (usually two inlet valves and two exhaust valves). Gives improved engine efficiency due to improved fuel/air mixture and exhaust gas flow in the *combustion chambers*.
Vee engine – An engine design in which the *cylinders* are set in two banks forming a 'V' when viewed from one end. A V8 for example consists of two rows of four cylinders each.
Venturi – A streamlined restriction in the *carburettor* throttle bore which causes a low pressure to occur; this sucks fuel into the air stream to form a vapour suitable for combustion.
Viscosity – A term used to describe the resistance of a fluid to flow. When associated with lubricating oil, it's given an *SAE* number, 10 being a very light oil and 140 being a very heavy oil.
Voltage regulator – A device which regulates the *alternator* output to a predetermined level. On most alternators the voltage regulator is an integral part of the alternator, and regulates the charging current as well as the voltage.

W

Wankel engine – A rotary engine which has a triangular shaped rotor which performs the function of the *pistons* in a conventional engine, and rotates in a housing shaped approximately like a broad-waisted figure of eight. Very few cars use this type of engine.
Wheel balancing – Adding small weights to the rim of a wheel so that there are no out-of-balance forces when the wheel rotates.
Wishbone – An 'A'-shaped *suspension* component, pivoted at the base of the 'A' and carrying a wheel at the apex. Normally mounted close to the horizontal.
Worm and nut steering – A steering system where the lower end of the steering column has a coarse screw thread on which a nut runs. The nut is attached to a spindle which carries the drop arm which, in turn, moves the steering linkage.

A comprehensive network of local radio stations now exists throughout the UK.

Most of these radio stations provide regularly updated reports on traffic flow and road conditions, which can be of great help to drivers. Listening to traffic reports will help you to avoid the inevitable 'jams' which occur during every day driving.

Some car radio/cassette players are now equipped with a 'Radio Data System' (RDS) which will automatically tune into special traffic information signals, interrupting normal radio or tape listening when bulletins are broadcast. An RDS-equipped radio may prove to be a worthwhile investment if you travel by car regularly, particularly when driving on business trips.

The following table provides details of all the local radio frequencies throughout the UK, region-by-region. Where stations broadcast on FM/VHF and AM/MW, it's suggested that the FM/VHF frequency is used wherever possible, as this will usually give better reception.

LOCAL RADIO FREQUENCIES

AREA	FM/VHF	AM/MW
AVON		
Bath		
BBC Bristol	104.6	1548
GWR FM	103.0	–
Bristol		
BBC Bristol	94.9	1548
Brunel Classic Gold	–	1260
Galaxy Radio	97.2	–
GWR FM	96.3	–
BEDFORDSHIRE		
Bedford		
BBC Bedfordshire	95.5	1161
Chiltern Radio	96.9	–
SuperGold	–	792
Luton		
BBC Bedfordshire	103.8	630
Chiltern Radio	97.6	–
SuperGold	–	828
BERKSHIRE		
Reading		
210 Classic Gold Radio	–	1431
210 FM	97.0	–
BBC Berkshire	104.4	–
Wokingham		
BBC Berkshire	104.1	–
BIRMINGHAM		
Birmingham		
BBC WM	95.6	1458
BRMB FM	96.4	–
Buzz FM	102.4	–
Xtra AM	–	1152
BORDERS		
Eyemouth		
Radio Borders	103.4	–
Peebles		
Radio Borders	103.1	–
Selkirk		
Radio Borders	96.8	–
BUCKINGHAMSHIRE		
Milton Keynes		
BBC Bedfordshire	104.5	630
Horizon Radio	103.3	–

AREA	FM/VHF	AM/MW
CAMBRIDGESHIRE		
Cambridge		
BBC Cambridge	96.0	1026
CN FM	103.0	–
Peterborough		
BBC Cambridge	95.7	1447
Hereward Radio	102.7	1332
CHANNEL ISLANDS		
Guernsey		
BBC Guernsey	93.2	1116
Jersey		
BBC Jersey	88.8	1026
CHESHIRE		
Echo 96	96.4	–
Chester		
BBC Merseyside	95.8	1485
Congleton		
Signal Cheshire	104.9	–
Macclesfield		
BBC Stoke	94.6	1503
Warrington		
BBC Merseyside	95.8	1485
CLEVELAND		
Middlesbrough		
BBC Cleveland	95.0	1548
TFM	96.6	–
GNR	–	1170
CLWYD		
Prestatyn		
BBC Cymru/Wales	94.2	882
Rhyl		
BBC Cymru/Wales	94.2	882
Wrexham		
BBC Cymru/Wales	93.3	657
MFM 1034	97.1/103.4	–
Marcher Gold	–	1260
CORNWALL		
Isles of Scilly		
BBC Cornwall	96.0	–
Liskeard		
BBC Cornwall	95.2	657
Redruth		
BBC Cornwall	103.9	630
CUMBRIA		
Barrow-in-Furness		
BBC Furness	96.1	837
Kendal		
BBC Cumbria	95.2	–
Windermere		
BBC Cumbria	104.2	–
Workington		
BBC Cumbria	95.6	1458

AREA	FM/VHF	AM/MW
DERBYSHIRE		
Chesterfield		
BBC Sheffield	94.7	1035
Derby		
BBC Derby	94.2	1116
Trent FM	102.8	–
Gem AM	–	945
Matlock		
BBC Derby	95.3	–
DEVON		
Barnstaple		
BBC Devon	94.8	801
Exeter		
BBC Devon	95.8	990
DevonAir	97.0	666/954
South West	103.0	–
Okehampton		
BBC Devon	96.0	801
Plymouth Area		
BBC Devon	103.4	855
Plymouth Sound	97.0	1152
Tavistock		
Radio in Tavistock	96.6	–
Torbay		
BBC Devon	103.4	1458
DevonAir	96.4	666/954
DORSET		
Poole		
BBC Solent	96.1	1359
Bournemouth		
BBC Solent	96.1	1359
2CR Classic Gold		828
2CR FM	102.3	–
Weymouth		
BBC Solent	96.1	–
DUMFRIES & GALLOWAY		
Dumfries		
BBC Scotland/Solway	94.7	585
SW Sound	97.2	–
Stranraer		
BBC Scotland	94.7	–
DURHAM		
Darlington		
BBC Newcastle	95.4	–
Durham		
BBC Newcastle	95.4	1458
DYFED		
Aberystwyth		
BBC Cymru/Wales	93.1	882

AREA	FM/VHF	AM/MW
EAST SUSSEX		
Brighton		
BBC Sussex	95.3	1485
South Coast Radio	–	1323
Southern Sound Classic Hits	103.5	–
Eastbourne		
BBC Sussex	104.5	1161
Southern Sound Classic Hits	102.4	–
Hastings		
Southern Sound Classic Hits	97.5	–
Newhaven		
Southern Sound Classic Hits	96.9	–
ESSEX		
Basildon		
BBC Essex	95.3	765
Chelmsford		
BBC Essex	103.5	765
Breeze AM		1431/1359
Essex Radio	102.6	–
Colchester		
BBC Essex	103.5	729
Mellow 1557	–	1557
Harlow		
BBC Essex	–	765
Manningtree		
BBC Suffolk	103.9	–
Southend-on-Sea		
BBC Essex	95.3	1530
Breeze AM		1431/1359
Essex Radio	96.3	–
FIFE		
Kirkcaldy		
BBC Scotland	94.3	810
GLOUCESTERSHIRE		
Gloucester		
BBC Gloucester	104.7	603
Severn Sound	102.4	
Three Counties Radio		774
Stroud		
BBC Gloucester	95.0	603
Severn Sound	103.0	–
GRAMPIAN		
Aberdeen		
BBC Scotland/Aberdeen	93.1	990
North Sound	96.9	1035
Stirling		
Central FM	96.7	–

AREA	FM/VHF	AM/MW
GWENT		
Newport		
Red Dragon Radio	97.4	–
Touch AM		1305
GWYNEDD		
Anglesey		
BBC Cymru/Wales	94.2	882
HAMPSHIRE		
Andover		
210 FM	102.9	–
Basingstoke		
210 FM	102.9	–
BBC Berkshire	104.1	–
Isle of Wight		
BBC Solent	96.1	999/1359
Isle of Wight Radio	–	1242
Portsmouth		
BBC Solent	96.1	999
Oceansound	97.5	–
South Coast Radio	–	1170
Southampton		
BBC Solent	96.1	999
Oceansound	97.5	–
Power FM	103.2	–
South Coast Radio	–	1557
Winchester		
Oceansound	96.7	–
Power FM	103.2	
HEREFORDSHIRE		
Hereford		
BBC Here/Worc	94.7	819
Radio Wyvern	97.6	954
HERTFORDSHIRE		
Watford		
BBC GLR	94.9	1458
St Albans		
BBC Bedfordshire	103.8	630
Stevenage		
BBC Bedfordshire	103.8	630
HIGHLAND		
Inverness		
BBC Scotland	94.0	810
Moray Firth Radio	97.4	1107
Fort William		
BBC Scotland	93.7	–
Oban		
BBC Scotland	93.3	–
Thurso		
BBC Scotland	94.5	–

AREA	FM/VHF	AM/MW
HUMBERSIDE		
Kingston upon Hull Area		
BBC Humberside	95.9	1485
Classic Gold	–	1161
Viking FM	96.9	–
ISLE OF MAN		
Douglas		
Manx Radio	97.2	1368
Snaefell		
Manx Radio	89.0	–
KENT		
Ashford		
Invicta FM	96.1	–
Canterbury		
BBC Kent	104.2	774
Invicta FM	102.8	–
Dover		
BBC Kent	104.2	774
Invicta FM	97.0	–
East Kent		
Coast Classics	–	603
Folkestone		
BBC Kent	104.2	774
Maidstone & Medway		
BBC Kent	96.7	1035
Coast Classics	–	1242
Invicta FM	103.1	–
Royal Tunbridge Wells		
BBC Kent	96.7	1602
Thanet		
Invicta FM	95.9	–
LANCASHIRE		
Burnley Area		
BBC Lancashire	95.5	855
Blackpool		
BBC Lancashire	103.9	855
Red Rose Gold	–	999
Red Rose Rock FM	97.4	–
Lancaster		
BBC Lancashire	104.5	1557
Preston		
BBC Lancashire	103.9	855
Red Rose Gold	–	999
Red Rose Rock FM	97.4	–
LEICESTERSHIRE		
Leicester		
BBC Leicester	104.9	837
Gem AM	–	1260
Sound FM	103.2	–

AREA	FM/VHF	AM/MW
LINCOLNSHIRE		
Lincoln Area		
BBC Lincoln	94.9	1368
Lincs FM	102.2	–
LONDON		
Brixton		
Choice FM	96.9	–
Ealing		
Sunrise Radio	–	1413
Greater London		
BBC GLR	94.9	1458
Capital FM	95.8	–
Capital Gold	–	1548
Jazz FM	102.2	–
Kiss FM	100.0	–
LBC Newstalk	97.3	–
Melody Radio	104.9	–
Spectrum Int. Radio	–	558/990
Haringey		
London Greek Radio	103.3	–
WNK	103.3	–
Thamesmead		
London Talkback Radio	–	1152
RTM	103.8	–
Southall		
Sunrise Radio	–	1413
LOTHIAN		
Bathgate		
Radio Forth	97.3	–
Dunfermline		
BBC Scotland	94.3	810
Edinburgh		
BBC Scotland	94.3	810
Max AM	–	1548
Radio Forth	97.3/97.6	–
MANCHESTER		
Manchester		
BBC GMR	95.1	1458
KFM	104.9	–
Piccadilly Gold	–	1152
Piccadilly Key 103	103.0	–
Sunset Radio	102.0	–
MERSEYSIDE		
Liverpool		
BBC Mersey	95.8	1485
City Talk	–	1548
Radio City	96.7	–
MID GLAMORGAN		
Aberdare		
BBC Cymru/Wales	93.6	882
Merthyr Tydfil		
BBC Cymru/Wales	96.8	882
MIDDLESEX		
Hounslow		
Sunrise Radio	–	1413
NORFOLK		
King's Lynn		
BBC Norfolk	104.4	873
Norwich Area		
BBC Norfolk	95.1	855
Broadland	102.4	1152
NORTHAMPTONSHIRE		
Kettering		
KCBC	–	1530
Northampton		
BBC Northampton	104.2	1107
Northants	96.6	–
SuperGold	–	1557
NORTHUMBERLAND		
Berwick-upon-Tweed		
BBC Newcastle	96.0	–
Radio Borders	97.5	–
NOTTINGHAM		
Nottingham		
BBC Nottingham	103.8	1521
Trent FM	96.2/96.5	–
Gem AM	–	999
ORKNEY		
Kirkwall		
BBC Scotland	93.7	–
OXFORDSHIRE		
Banbury		
Fox FM	97.4	–
Oxford		
BBC Oxford	95.2	1485
Fox FM	102.6	–
POWYS		
Welshpool		
BBC Cymru/Wales	94.0	882
SHETLAND		
Lerwick		
BBC Scotland	92.7	–
SIBC	96.2	–
SHROPSHIRE		
Ludlow		
BBC Shropshire	95.0	1584
RFM	97.6	–
Shrewsbury		
BBC Shropshire	96.0	–
Beacon Radio	97.2/103.1	
WABC	–	1017
Telford		
BBC Shropshire	96.0	756
Beacon Radio	97.2/103.1	
WABC	–	1017
SOMERSET		
Mendip Area		
Orchard FM	102.6	–
Wells		
BBC Bristol	95.5	1548
SOUTH GLAMORGAN		
Cardiff		
BBC Cymru/Wales	96.8	882
Red Dragon Radio	103.2	–
Touch AM		1359
STAFFORDSHIRE		
Stafford		
Echo 96	96.9	–
Stoke on Trent		
BBC Stoke	94.6	1503
Signal Radio	102.6	1170
STRATHCLYDE		
Ayr		
West Sound	96.7	1035
Girvan		
West Sound	97.5	–
Glasgow		
BBC Scotland	94.3	810
Clyde 1	102.5	–
Clyde 2	–	1152
Radio Clyde FM	97.3	
Greenock		
BBC Scotland	94.3	810
Kilmarnock		
BBC Scotland	93.9	810
SUFFOLK		
Bury St. Edmunds		
Saxon Radio	96.4	1251
Great Barton		
BBC Suffolk	104.6	–
Ipswich		
BBC Suffolk	103.9	–
Radio Orwell	97.1	1170

AREA	FM/VHF	AM/MW
SURREY		
Guildford		
BBC Surrey	104.6	–
Delta Radio	97.1	–
First Gold Radio	–	1476
Premier Radio	96.4	–
Reigate		
Radio Mercury	102.7	1521
TAYSIDE		
Dundee		
Radio Tay	102.8	1161
BBC Scotland	92.7	810
Perth		
Radio Tay	96.4	1584
TYNE & WEAR		
Fenham		
BBC Newcastle	104.4	–
Metro FM	103.0	–
Newcastle Area		
BBC Newcastle	95.4	1458
GNR	–	1152
Metro FM	97.1	–
Sunderland		
Wear FM	103.4	–
ULSTER		
Belfast		
BBC Ulster	94.5	1341
Classic Trax BCR	96.7	–
Cool FM	97.4	–
Downtown Radio	102.6	–
Enniskillen		
BBC Ulster	93.8	873
Downtown Radio	96.6	–
Limavady		
Downtown Radio	96.4	–
Londonderry		
BBC Foyle	93.1	792
Downtown Radio	102.4	–
WARWICKSHIRE		
Leamington Spa		
Mercia FM	102.9	–

AREA	FM/VHF	AM/MW
WEST GLAMORGAN		
Swansea		
BBC Cymru/Wales	93.9	882
Swansea Sound	96.4	1170
WEST MIDLANDS		
Coventry		
BBC CWR	94.8	–
Mercia FM	97.0	–
Radio Harmony	102.6	–
Xtra AM	–	1359
Sutton Coldfield		
BBC Derby	104.5	–
Beacon Radio	97.2/103.1	–
Wolverhampton		
BBC WM	95.6	828
Beacon Radio	97.2/103.1	–
WABC	–	990
WEST SUSSEX		
Crawley		
Radio Mercury	102.7	1521
Horsham		
BBC Sussex	95.1	1368
Worthing		
BBC Sussex	95.3	1485
WESTERN ISLES		
Stornoway (Lewis)		
BBC Scotland	94.2	–
WILTSHIRE		
Chippenham		
BBC Wiltshire Sound	104.3	–
Marlborough		
GWR FM	96.5	–
Salisbury		
BBC Wiltshire Sound	103.5	–
Swindon		
BBC Wiltshire Sound	103.6	1368
Brunel Classic Gold	–	1161
GWR FM	97.2	–
West Wilts		
Brunel Classic Gold	–	936
GWR FM	102.2	–
WORCESTERSHIRE		
Worcester		
BBC Here/Worc	104.0	738
Radio Wyvern	102.8	1530

AREA	FM/VHF	AM/MW
YORKSHIRE NORTH		
Northallerton		
BBC York	104.3	–
Scarborough		
BBC York	95.5	1260
Whitby		
BBC Cleveland	95.8	–
York		
BBC York	103.7	666
YORKSHIRE SOUTH		
Barnsley		
Classic Gold	–	1305
Hallam	102.9	–
Doncaster		
BBC Sheffield	104.1	1035
Classic Gold	–	990
Hallam	103.4	–
Rotherham		
BBC Sheffield	88.6	1035
Hallam	96.1	–
Sheffield		
BBC Sheffield	88.6	1035
Classic Gold	–	1548
Hallam	97.4	–
YORKSHIRE WEST		
Bradford		
Classic Gold	–	1278
Pennine	97.5	–
Sunrise FM	103.2	–
Halifax		
Classic Gold	–	1530
Pennine	102.5	–
Huddersfield		
Classic Gold	–	1530
Pennine	102.5	–
Leeds		
Aire FM	96.3	–
BBC Leeds	92.4	774
Magic 828	–	828

Length (distance)

Inches *(in)*	x 25.4	= Millimetres *(mm)*	x 0.0394	= Inches *(in)*
Feet *(ft)*	x 0.305	= Metres *(m)*	x 3.281	= Feet *(ft)*
Miles	x 1.609	= Kilometres *(km)*	x 0.621	= Miles

Volume (capacity)

Cubic inches *(cu in; in³)*	x 16.387	= Cubic centimetres *(cc; cm³)*	x 0.061	= Cubic inches *(cu in; in³)*
Pints *(pt)*	x 0.568	= Litres *(l)*	x 1.76	= Pints *(pt)*
Quarts *(qt)*	x 1.137	= Litres *(l)*	x 0.88	= Quarts *(qt)*
Gallons *(gal)*	x 4.546	= Litres *(l)*	x 0.22	= Gallons *(gal)*

Mass (weight)

Ounces *(oz)*	x 28.35	= Grams *(g)*	x 0.035	= Ounces *(oz)*
Pounds *(lb)*	x 0.454	= Kilograms *(kg)*	x 2.205	= Pounds *(lb)*

Force

Pounds-force *(lbf; lb)*	x 4.448	= Newtons *(N)*	x 0.225	= Pounds-force *(lbf; lb)*
Newtons *(N)*	x 0.1	= Kilograms-force *(kgf; kg)*	x 9.81	= Newtons *(N)*

Pressure

Pounds-force per square inch *(psi; lbf/in²; lb/in²)*	x 0.070	= Kilograms-force per square centimetre *(kgf/cm²; kg/cm²)*	x 14.223	= Pounds-force per square inch *(psi; lbf/in²; lb/in²)*
Pounds-force per square inch *(psi; lbf/in²; lb/in²)*	x 0.068	= Atmospheres *(atm)*	x 14.696	= Pounds-force per square inch *(psi; lbf/in²; lb/in²)*
Pounds-force per square inch *(psi; lbf/in²; lb/in²)*	x 0.069	= Bars	x 14.5	= Pounds-force per square inch *(psi; lbf/in²; lb/in²)*

Torque (moment of force)

Pounds-force feet *(lbf ft; lb ft)*	x 0.138	= Kilograms-force metres *(kgf m; kg m)*	x 7.233	= Pounds-force feet *(lbf ft; lb ft)*
Pounds-force feet *(lbf ft; lb ft)*	x 1.356	= Newton metres *(Nm)*	x 0.738	= Pounds-force feet *(lbf ft; lb ft)*
Newton metres *(Nm)*	x 0.102	= Kilograms-force metres *(kgf m; kg m)*	x 9.804	= Newton metres *(Nm)*

Power

Horsepower *(hp)*	x 0.745	= Kilowatts *(kW)*	x 1.3	= Horsepower *(hp)*

Velocity (speed)

Miles per hour *(miles/hr; mph)*	x 1.609	= Kilometres per hour *(km/h; kph)*	x 0.621	= Miles per hour *(miles/hr; mph)*

Fuel consumption*

Miles per gallon *(mpg)*	x 0.354	= Kilometres per litre *(km/l)*	x 2.825	= Miles per gallon *(mpg)*

Temperature

Degrees Fahrenheit = (°C x 1.8) + 32

Degrees Celsius (Degrees Centigrade; °C) = (°F – 32) x 0.56

** Note: It is common practice to convert from miles per gallon (mpg) to litres/100 kilometres (l/100 km), where mpg x l/100 km = 282*

MILES

	LONDON	Aberdeen	Aberystwyth	Birmingham	Bournemouth	Brighton	Bristol	Cambridge	Cardiff	Carlisle	Chester	Derby	Dover	Edinburgh	Exeter	Fishguard	Fort William	Glasgow	Harwich	Holyhead
LONDON		806	341	179	169	89	183	84	249	484	298	198	117	608	275	422	818	634	114	420
Aberdeen	501		724	658	903	895	789	742	792	349	571	634	924	198	911	795	254	241	848	692
Aberystwyth	212	450		193	328	425	206	354	177	375	153	222	459	526	325	90	708	525	460	182
Birmingham	111	409	120		237	267	132	161	163	309	117	63	296	460	259	283	642	459	267	246
Bournemouth	105	561	204	147		148	124	251	195	554	351	306	278	705	132	375	887	703	283	463
Brighton	55	556	264	166	92		217	180	288	573	386	286	130	697	275	468	906	723	203	509
Bristol	114	490	128	82	77	135		233	71	439	237	195	314	591	121	251	722	589	314	340
Cambridge	52	461	220	100	156	112	145		290	417	269	155	204	544	351	436	750	566	106	391
Cardiff	155	492	110	101	121	179	44	180		443	237	225	385	587	192	180	776	608	359	341
Carlisle	301	217	233	192	344	356	273	259	275		222	304	602	151	571	444	333	150	523	341
Chester	185	355	95	73	218	240	147	167	147	138		114	415	373	357	243	555	372	375	137
Derby	123	394	138	39	190	178	121	96	140	189	71		315	436	315	312	637	454	261	251
Dover	73	574	285	184	173	81	195	127	239	374	258	196		726	391	539	935	752	203	538
Edinburgh	378	123	327	286	438	433	367	338	365	94	232	271	451		723	616	233	72	650	489
Exeter	171	566	202	161	82	171	75	218	119	355	222	196	243	449		372	904	721	389	460
Fishguard	262	494	56	176	233	291	156	271	112	276	151	194	335	383	231		777	594	533	272
Fort William	508	158	440	399	551	563	480	466	482	207	345	396	581	145	562	483		183	856	671
Glasgow	394	150	326	285	437	449	366	352	378	93	231	282	467	45	448	369	114		673	488
Harwich	71	527	286	166	176	126	195	66	223	325	233	162	126	404	242	331	532	418		497
Holyhead	261	430	113	153	288	316	211	243	212	212	85	156	334	304	286	169	417	303	309	
Hull	168	348	229	143	255	223	226	124	244	155	132	98	251	225	301	285	362	250	181	217
Inverness	537	105	482	445	594	590	528	497	524	253	387	430	601	159	603	529	66	176	563	463
Leeds	190	322	173	111	261	251	206	148	212	115	78	74	269	199	268	229	322	210	214	163
Leicester	98	417	151	39	166	153	112	68	139	208	94	28	171	294	187	207	415	301	134	183
Lincoln	136	382	186	85	217	191	163	86	190	180	123	51	213	259	238	245	387	273	152	208
Liverpool	205	334	104	94	235	260	164	184	164	116	17	88	278	210	239	160	323	211	250	94
Manchester	192	332	133	81	228	247	167	154	172	115	38	58	265	209	242	189	322	210	220	123
Newcastle	271	230	270	209	348	326	295	231	310	58	175	164	344	107	370	326	257	143	297	260
Norwich	107	487	267	155	212	162	221	60	237	287	210	139	169	364	277	331	494	380	63	299
Nottingham	123	388	154	50	189	178	132	84	152	185	87	16	196	265	207	210	392	278	150	172
Oxford	56	464	157	62	92	99	66	79	107	256	128	98	129	341	139	207	461	340	134	202
Penzance	282	683	313	268	193	269	186	329	230	466	333	307	354	560	111	342	673	559	353	397
Plymouth	213	614	244	199	124	213	117	260	161	397	264	238	285	491	42	273	604	490	284	328
Preston	216	302	144	105	256	271	185	198	190	85	49	100	289	179	266	200	292	180	264	125
Sheffield	162	358	158	75	227	217	163	124	176	145	78	36	235	235	248	214	352	240	190	163
Southampton	79	530	202	128	32	60	75	136	119	322	194	164	149	416	111	231	520	415	150	283
Stranraer	410	235	342	301	453	465	382	368	384	109	247	298	483	133	464	385	199	85	434	321
Swansea	196	506	76	126	162	220	85	216	41	289	151	165	280	383	160	71	496	382	264	189
Worcester	114	428	98	26	131	168	60	116	75	211	88	65	187	305	135	155	418	304	182	152
York	196	309	197	134	273	251	215	157	242	116	102	89	269	186	290	253	323	211	223	187

KILOMETRES

	Hull	Inverness	Leeds	Leicester	Lincoln	Liverpool	Manchester	Newcastle	Norwich	Nottingham	Oxford	Penzance	Plymouth	Preston	Sheffield	Southampton	Stranraer	Swansea	Worcester	York
LONDON	270	864	306	158	219	330	309	436	172	198	90	454	343	348	261	127	660	315	183	315
Aberdeen	560	169	518	671	615	538	534	370	784	624	747	1099	988	486	576	853	378	814	689	497
Aberystwyth	369	776	278	243	304	167	214	435	430	248	253	504	393	232	254	325	550	122	158	317
Birmingham	230	716	179	63	137	151	130	336	249	80	100	431	320	169	121	206	484	203	42	216
Bournemouth	410	956	420	267	349	378	367	560	341	304	148	311	200	412	365	51	729	261	211	439
Brighton	359	950	404	246	307	418	398	525	261	286	159	433	343	436	349	97	748	354	270	404
Bristol	364	850	332	180	262	264	269	475	356	212	106	299	188	298	262	121	615	137	97	346
Cambridge	200	800	238	109	138	296	248	372	97	135	127	529	418	319	200	219	592	348	187	253
Cardiff	393	843	341	224	306	264	277	499	381	245	172	370	259	306	283	192	618	66	121	389
Carlisle	249	407	185	335	290	187	185	93	462	298	412	750	639	137	233	518	175	465	340	187
Chester	212	623	126	151	198	27	61	282	338	140	206	536	425	79	126	312	398	243	142	164
Derby	158	692	119	45	82	142	93	264	224	26	158	494	383	161	58	264	480	266	105	143
Dover	404	967	433	275	343	447	426	554	272	315	208	570	459	465	378	240	777	451	301	433
Edinburgh	362	256	320	473	417	338	336	172	586	426	549	901	790	288	378	669	214	616	491	299
Exeter	484	970	431	301	383	385	389	595	446	333	224	179	68	428	399	179	747	257	217	467
Fishguard	459	851	369	333	394	257	304	525	533	338	333	550	439	322	344	372	620	114	249	407
Fort William	583	106	518	668	623	520	518	414	795	631	742	1083	972	470	566	837	320	798	673	520
Glasgow	402	283	338	484	439	340	338	230	612	447	547	900	789	290	386	668	137	615	489	340
Harwich	291	906	344	216	245	402	354	478	101	241	216	568	457	425	306	241	698	429	293	359
Holyhead	349	745	262	295	335	151	198	418	481	277	325	639	528	201	262	455	517	304	245	301
Hull		615	90	145	61	196	151	190	240	148	262	663	552	180	103	412	425	433	272	63
Inverness	382		579	729	673	605	604	428	842	682	816	1159	1049	555	634	925	417	872	747	552
Leeds	56	358		158	108	119	64	148	277	116	272	613	502	90	58	378	360	369	220	39
Leicester	90	453	98		82	187	138	301	187	42	119	480	369	206	105	225	510	282	109	174
Lincoln	38	418	67	51		192	137	245	169	56	201	562	451	185	72	307	465	340	192	124
Liverpool	122	376	74	116	119		55	253	360	167	251	563	452	50	119	357	362	270	169	158
Manchester	94	375	40	86	85	34		212	306	119	230	568	457	48	64	325	360	295	164	103
Newcastle	118	266	92	187	152	157	132		414	254	412	774	663	203	206	518	262	517	378	127
Norwich	149	523	174	116	105	224	190	257		198	224	626	515	354	241	299	637	439	283	293
Nottingham	92	424	72	26	35	104	74	158	123		156	512	401	167	64	262	473	283	122	134
Oxford	163	507	169	74	125	156	143	256	139	97		402	291	272	217	106	587	230	95	291
Penzance	412	720	381	298	349	350	353	481	389	318	250		126	597	560	357	925	436	396	645
Plymouth	343	652	312	229	280	281	284	412	320	249	181	78		486	447	246	814	325	285	534
Preston	112	345	56	128	115	31	30	126	220	104	169	371	302		113	375	312	322	203	126
Sheffield	64	394	36	65	45	74	40	128	150	40	135	348	278	70		323	490	336	163	90
Southampton	256	575	235	140	191	222	202	322	186	163	66	222	153	233	201		679	257	193	398
Stranraer	264	259	224	317	289	225	224	163	396	294	365	575	506	194	254	422		641	515	362
Swansea	269	542	229	175	211	168	183	321	273	176	143	271	202	200	209	160	398		161	418
Worcester	169	464	137	68	119	105	102	235	176	76	59	246	177	126	101	120	320	100		257
York	39	343	24	108	77	98	64	79	182	83	181	401	332	78	56	247	225	260	160	

A

About this Handbook 5
ABS (anti-lock braking system)
warning light 22, 23
Accidents 35
Dealing with 35
First aid 36
Recording details of 38
Requirements of the law 38
Air cleaner air intake temperature
control 107, 108
Air cleaner element, renewal ....105 to 107
Air conditioning controls 27
Alcohol and driving 54
Alternator drivebelt, checking 100, 102, 114
Asbestos, precautions concerning 87
Automatic transmission 25, 104
Fluid level 82, 101
Auxiliary warning system 23

B

Bad-weather driving 50
Battery:
Electrolyte level 78
Precautions 87
Starting when flat 47
Terminals 114
Bodywork, checks and
maintenance 99, 115, 117, 118
Bonnet release lever 28, 102
Boot light 126
Boot/tailgate release 102
Brake fluid level 71, 76
Brake fluid low level warning light 22, 97
Brake pad wear warning light 23
Braking system, checks 102, 104
Breakdowns 43
Emergency kit 48
Broken windscreen, dealing with 42
Bulb failure warning light 24
Bulbs, renewal 119
Boot light 126
Courtesy light 125
Direction indicator repeater light 124
Direction indicator (front) 121, 122
Driving light 122
Foglight (front) 123
Glovebox light 126
Headlight 120
Luggage compartment light 126
Map reading light 126
Number plate light (rear) 125
Rear lights 124, 125
Sidelight (front) 121
Bump starting 47
Buying and selling 9

C

Capacities, lubricants and fluids 67
Car crime prevention 65
Car jargon 137
Caravan, towing ....52 to 54
Central door locking 33
Changing a wheel 44
Child safety 57
Cigarette lighter 29
Clock 19
Clutch 104
Coil (ignition) ....108 to 111
Computer 21, 22
Controls and equipment 15
Conversion factors 152, 153
Coolant:
Checking strength 114
Level 71, 75
Renewal 112
Coolant level warning light 23
Coolant pump drivebelt (OHC engines) 100
Cooling system, flushing 113
Courtesy light 125
Crankcase emission valve
(OHC and V6 engines) 104
Crime prevention 65

D

Dimensions and weights 13
Direction indicator (front),
bulb renewal 121, 122
Direction indicator repeater light,
bulb renewal 124
Direction indicator warning light 23
Distance tables 154, 155
Distributor, check and clean ....108 to 111
Door locks 32, 33, 102
Door mirror switch (electric) 27, 28
Door-open warning light 24
Driveshafts 104

Driving:
abroad 59
in bad weather 50
on motorways 51
safety 49
Driving light, bulb renewal 122

E

Economical driving 63
Econolights 23
Electric windows 29
Electric door mirror switch 27, 28
Electrical system, checks 71, 99, 102, 114
Electricity (mains) and electrical equipment, precautions concerning 88
Emergencies 35
Engine compartment layouts ....72 to 74, 90 to 94
Engine oil, disposal of 88
Engine oil and filter, renewal 95
Engine oil filler cap, cleaning 97
Engine performance, checking 104
Engine types, identifying 89
Environmental protection 63, 64, 87, 88
Equipment 15

F

Fanbelt 100, 102, 114
Fault finding 131
Brakes 134
Electrics 135, 136
Engine ....131 to 133
Gearbox, transmission and clutch 133
Suspension and steering 134, 135
Fire hazards, precautions concerning 87
Fire in vehicle, dealing with 42
First aid 36
Foglight switches 27
Front foglight switch 27
Front seat adjustment 30
Fuel computer 21
Fuel gauge 19
Fuel level warning light 23
Fuel system, checks and adjustments ....82, 97, 105 to 107
Fumes, precautions 87
Fuses, renewal 126

G

Gearbox (manual) 24, 104
Glossary 137
Glovebox light, bulb renewal 126
Graphic warning system 23

H

Handbrake 102
Handbrake warning light 22
Hazard flasher (warning) switch 27
Head restraints 30
Headlight, bulb renewal 120
Headlight main beam warning light 23
Heated front seat switch 28
Heated rear window switch 27
Heated windscreen switch 28
Heating and ventilation controls 26
Hinges 102
History of Sierra range 7
Horn 25, 71, 79
Hoses 95, 112, 114
HT leads ....108 to 111

I

Ice warning light 23
Identifying engine types 89
Ignition HT voltage, precautions concerning 88
Ignition switch/steering lock 24
Ignition warning light 22
Instrument illumination control 28
Instrument panel layouts ....15 to 18
Instrumentation 19, 103
Interior 117

J

Jacking and vehicle support 88, 89
Jacking and wheel changing 44
Jargon 137
Jump lead starting 47

L

Leaks (fluid), checking for 71, 79
Lights, checking operation 71, 79, 114
Lights switch (exterior lights) 25
Local radio frequencies ......147 to 151
Locks 32, 33, 102
Loudspeaker balance control 28
Lubricants and fluids 68
Luggage compartment light, bulb renewal .126

M

Mains electricity, precautions 88
Maintenance schedule ......see Service schedule
MOT, preparing for 129
  Brakes 129
  General 130
  Lights 129
  Seatbelts 130
  Steering and suspension 129
  Tyres and wheels 130
Motorway breakdowns 44
Motorway driving 51
Multi-function switches 25

N

Number plate light (rear), bulb renewal 125

O

Oil and filter, renewal 95
Oil, disposal of 88
Oil filler cap, cleaning 97
Oil level warning light 23
Oil level 71, 74
Oil pressure warning light 22

P

Paintwork scratches 99, 118
Parcel shelf 32
Pipes 95
Plugs (spark) 100
Power steering fluid level 98
Power steering pump drivebelt 111
Push starting 47

R

Radio frequencies ......147 to 151
Radiator and hoses 112, 114
Rear lights, bulb renewal 124, 125
Rear foglight switch 27
Rear parcel shelf 32
Rear seats ......30 to 32
Rear view mirror (interior) 29
Rear window sun-blinds 32
Recommended lubricants and fluids 68
Reducing the cost of motoring 63
Regular checks 71
Relays, renewal 128
Remote boot/tailgate release 28
Rev. counter 19
Road testing 103
Roadwheel bolts 44, 98
Routine checks 71
Routine maintenance 71, 81

S

Safety (child) 57
Safety (driving) 49
Safety first! 86
Scratches 118
Seat adjustment 30
Seat belt warning light 23
Seat belts 29, 30, 99
Security 65
Selling 11
Service schedule ......82 to 85
  250 miles (400 km) or weekly 71, 95
  6000 miles (10 000 km) or 6 months ......95 to 100
  12 000 miles (20 000 km) or 12 months ......100 to 104
  24 000 miles (40 000 km) or 2 years ......104 to 111
  36 000 miles (60 000 km) 111
  36 000 miles (60 000 km) or 2 years ......112 to 114
  36 000 miles (60 000 km) or 3 years 114
  Seasonal servicing 114
Service specifications 67, 68
Service tasks 95
Servicing 81
Sidelight (front), bulb renewal 121
Sidelight/tail light warning light 124, 125
Sierra family 7

Skid control 54
Spare parts 89
Spark plug specifications 67
Spark plugs, renewal 100
Speaker balance control 28
Specifications 67, 68
Speedometer 19
Starting a car with a flat battery 47
Steering, checks 98, 104
Steering column switches 25
Steering lock 24
Sunroof 29
Suspension, checks 104
Switches:
  Electric door mirror 27, 28
  Front foglight 27
  Hazard flasher (warning) 27
  Heated rear window 27
  Heated windscreen 28
  Heated front seat 28
  Horn 25
  Ignition 24
  Lights 25
  Multi-function 25
  Rear foglight 27
  Steering column 25
  Tailgate wash/wipe 26
  Windscreen wash/wipe 25, 26

T

Tachometer 19
Tail lights, bulb renewal 124, 125
Tailgate wash/wipe switch 26
Tailgate release 102
Temperature gauge 19
Throttle linkage 97
Tools 116
Tow starting 47
Towing ....46, 52 to 54
Trailer, towing ....52 to 54
Trip computer 22
Troubleshooting ....see Fault finding
Tyre pressures 68, 76
Tyres, checking 76, 77, 115

U

Underbody, checking and cleaning 99, 115
Underbonnet views ....72 to 74, 90 to 94

V

Vandalism, dealing with 42
Ventilation controls 26

W

Warning lights 22
  ABS (anti-lock braking system) 22, 23
  Brake pad wear 23
  Brake fluid, low level 22
  Coolant level 23
  Direction indicator 23
  Econolights 23
  Fuel level 23
  Handbrake 22
  Headlight main beam 23
  Ignition 22
  Oil level 23
  Oil pressure 22
  Seat belt 23
  Washer fluid level 23
Wash/wipe switch 25, 26
Washer fluid level 71, 78, 114
Washer fluid level warning light 23
Weekly checks 71
Weights 13
Wheel bolts 44, 98
Wheel changing 44
Windows 102
Windscreen breakage, dealing with 42
Windscreen wash/wipe switch 25, 26
Wipers and washers, checking 71, 78, 114
Women drivers, advice to 55

Sierra Manuals

Like this driver's handbook, the above manuals have been written specifically for your Sierra. Most models are covered from 1982 to 1992, the only exceptions being those fitted with Diesel engines (for which a separate manual on Ford Diesel engines is produced) and the RS Cosworth.

The manuals are written in the same easy-to-follow manner as the driver's handbooks and enable you to tackle even major overhauls and repairs yourself. They will advise you on which jobs are suitable for the DIY mechanic and often suggest practical alternatives to the specialist tools that are so often recommended.

With the ever increasing cost of garage labour, think of the money you could save!

Haynes Manuals are available from most good motor accessory stores and bookshops or direct from the publisher.

Haynes THE BOOK ®

HAYNES

Haynes Publishing, Sparkford, Nr Yeovil, Somerset BA22 7JJ Telephone 0963 440635